Sadhan Kumar Ghosh · Ulhas '
Kåre Helge Karstensen

C000281314

Sustainable Management of Wastes Through Co-processing

 Springer

Sadhan Kumar Ghosh
Department of Mechanical Engineering
International Society of Waste
Management, Air and Water (ISWMAW),
Jadavpur University
Kolkata, West Bengal, India

Ulhas V. Parlikar
Ex. Geocycle India
Mumbai, India

Kåre Helge Karstensen
SINTEF
Oslo, Norway

ISBN 978-981-16-6075-7   ISBN 978-981-16-6073-3   (eBook)
https://doi.org/10.1007/978-981-16-6073-3

This Springer imprint is published by the registered company Springer Nature Singapore Pte Ltd.
The registered company address is: 152 Beach Road, #21-01/04 Gateway East, Singapore 189721,
Singapore

Sustainable Management of Wastes Through Co-processing

"Sustainable Management of Wastes through Co-processing"

is globally the first comprehensive book on Co-processing.

This book is being dedicated to

Mrs. Pranati Ghosh, beloved wife of Prof Sadhan K Ghosh,

Mrs. Rashmi Parlikar, beloved wife of Mr. Ulhas V. Parlikar,

Mrs. Eilen Henningsen, beloved wife of Dr. Kåre Helge. Karstensen,

who have been constant support & inspiration while writing the book

and

to the researchers, academia, policy makers, engineers, designers, cement plant operators, municipal administration, cement industries involved in co-processing, cement manufacturers Associations, other related stakeholders, waste management agencies, AFR traders, pre-processing agencies, analytical testing laboratories who all participate in the act of co-processing and also to the whole of the civil society who get benefited by the act of co-processing

Preface

Alternative fuels and raw materials (AFR) from waste can play an important role in contributing towards reducing the use of fossil fuel and costs while conserving natural resources, lowering global CO_2 emissions, and reducing the need for landfills. The use of AFR in resource and energy intensive industries is called co-processing. Emission reductions from co-processing AFs are dependent on the emissions factor and biomass content. Agricultural residues and biomass residues such as waste wood, rice husks, dried sewage sludge or animal meal and other fuels with high biogenic carbon content can be considered as carbon neutral. AFs derived from waste materials such as waste oil and non-recyclable plastics have varying emissions values, which are usually lower than traditional fossil fuels. Increasingly, fuels are used which contain both fossil and biogenic carbon, e.g., pre-treated industrial wastes, waste tires, or RDF from MSW which contains biogenic carbon.

Co-processing in cement kilns is a technology that is practiced globally on large scale for environmentally sound and ecologically sustaining management of wastes from agricultural, industrial, and municipal sources. Although co-processing may also be carried out in other energy & resource intensive industries like power generating plants, steel plants, refractory, lime, etc., it is widely practiced in cement plants due to versatility of the cement kiln operation in utilizing wastes.

Considerable amount of scientific and technological advancements have been put in place while developing and implementing this technology at the cement plant operational scales. This technology is in practice for about 40 years or so and has been recommended by Basel Convention for the sustainable management of Hazardous wastes and by the Stockholm Convention for the sustainable management of persistent organic pollutants (POPs). Co-processing promotes mitigation of the climate change impacts and also conservation of the natural capital in addition to building circular economy on large scale with a potential to provide large scale employment. Co-processing is an energy and material recovery process and addresses the issues related to sustainable management of industrial wastes. In cement kiln, the combustible portion of the waste gets used as fuel and the non-combustible portion gets used as the raw material leaving nothing to be disposed in landfill or any other option of treatment. The combustible portion replaces the fossil fuels and the non-combustible portion (Ash) replaces the raw materials such as Calcium, Aluminium, Silica, and Iron. Due to high temperature, long residence time and alkaline environment present in cement kilns, the environmental impact of the waste management in cement kilns is negligible.

Co-processing technology has been encouraged in some countries incorporating in national policies, strategies, and programmes. Co-processing technology has been included in the waste management rules notified by the government of India in the year 2016 as a preferred option for wastes management over the conventional options of incineration and landfill. The regulatory framework for co-processing and their subsequent amendments have been established in several countries, namely European countries, Australia, Brazil, China, India, Japan, South Africa, and United States. The understanding and awareness of the stakeholders belonging to the academic and other relevant sections is increasing at a slower rate which needs to be accelerated. Number of researchers and research publications on co-processing in a few countries in Asia, Europe, and USA is increasing though more research efforts have to be encouraged. To be more specific, research on co-processing and associated publications is found concentrated mainly in Austria, China, Germany, India, Japan, and Norway. The understanding of co-processing technology is found to be very limited within the stakeholders involved in various aspects, namely a) the implementation of the policy framework, b) design and engineering of the waste processing facilities to suit the co-processing operation, c) environmental consideration in implementing co-processing, and d) operation and management of the cement plant, quality and parameter controls, safety, etc. A huge potential of co- processing of waste exists worldwide. The stronger national legislation will make the co-processing widening the scopes in the countries.

Globally, co-processing in cement kilns is in practice for more than four decades for utilizing a large quantum of hazardous and non-hazardous waste. The cement kilns in Austria and Germany manage over 70% of their fuel requirement derived from co-processing AFRs. Almost entire quantum of organic hazardous waste generated in Norway is co-processed in the cement kilns. Cement kiln co-processing technology has the advantage that barring a few waste materials which are termed as "banned items", most of all the possible waste materials can be co-processed in a sustainable manner with 100% material recovery and 100% energy recovery. Co-processing is fully aligned to the concept of waste management hierarchy and circular economy. In the current requirement of resource efficiency improvement, co-processing has opportunity to play a substantially large role.

Municipal Solid Waste Incinerators with waste to energy (WtE) normally involves generation of heat and electricity. The conversion efficiency to electricity in WtE plants is poor and will usually not recover the construction costs. WtE-plants are expensive to build and operate; they represent an additional emission source and produce large amounts of residues (exit gas scrubbing residues, fly ash, bottom ash, etc.) that need to be treated and landfilled. Incineration of wet wastes in the rainy season is another challenge, which causes difficult burning conditions and often results in elevated emissions. Countries with cement industry may to a certain degree forego building expensive WtE's/incineration plants. The cement kilns are already in operation and may increase the waste treatment capacity significantly if integrated into the waste management strategy. They are usually cost-efficient and do not produce any residues that needs disposal. Cement kilns have proven to be effective means of recovering value from waste materials and co-processing in cement kilns is now an integral component in the spectrum of viable options for treating several waste categories, practised in developed countries for the last four decades.

Presently, the two cement plants in Norway replace more than 75% of its coal with waste and this has been the only treatment option for disposal of organic hazardous wastes in Norway for the last 30 years—a dedicated incinerator for hazardous wastes was never built. This practice has been cost-effective, resource-efficient, and environmentally sound compared to incineration. The energy utilization efficiency is much better than in an Incinerator with WtE—and no residues are produced, compared to around 30% in a WtE.

This book entitled *Sustainable Management of Wastes through Co-processing* is the first comprehensive book on Co-processing as there is possibly no book available in the world covering all aspects of co-processing. To the best of our understanding, no specific book could be found that is written on the subject of co-processing which could serve the need of all the relevant stakeholders related to sustainable management of wastes. There are a few guidelines brought out by different agencies such as UNEP, GIZ, HOLCIM, CPCB, WBCSD, and CSI on the subject of co-processing. A publication has been brought out by Dr. Dirk Lechtenberg and Dr. Hansjorg Diller with the title "Handbook of Alternative Fuels and Raw materials for cement and lime industries" which covers aspects related to different AFR materials.

This book consists of fifteen chapters divided into six sections given in the following table. The book covers all the aspects of co-processing of wastes in cement plants. It presents a collection of comprehensive relevant definitions relevant to co-processing taken from different sources, presents a status review of research carried out on co-processing, cement chemistry, theories and practices on cement manufacturing, pre-processing and co-processing, different guidelines, legislation, AFR, operations management and emission considerations, plants and equipment, Business Models, Case studies from significant co-processing operators in India, global scenario, Growth and Advocacy. The book has an appendix for search at the end. More than 100 references in each chapter.

The summary of contents in each of the chapters has been given in Chap. 1.

Section I   Introducing the subject

    1   Introduction
    2   Terms and Definition related to Co-processing of waste in cement kiln

Section II   Literature Review

    3   Status Review of Research on Co-Processing

Section III   Cement: Theory, Production Technology, and operations

A major gap in the knowledge on co-processing and its associated aspects has been observed among the policy makers, cement plant operators, engineers, designers, researchers, academia, and as a whole in the civil society while co-processing should be evolved as the best energy and materials recovery option with zero residue for waste management in cement kilns. Co-processing is an important technology for the sustainable management of hazardous and non-hazardous wastes derived out of Municipal, Industrial and Agricultural sources and it needs to be studied in universities, colleges, and schools as a course subject of waste management, Environmental Science, Environmental Engineering, Civil Engineering, Chemical Engineering, and Sustainability engineering, etc. Robust endeavour is essential to carry out quality research on co-processing and allied areas by the researchers and funding support for the research by the industry and the government. More international cooperation is required for strengthening the co-processing system, technology, strategies, and implementation. This book has been developed with an objective to fulfil these needs.

The book has been written in simple and lucid language focusing the target audience, namely researchers, academia, policy makers, cement plant operators, engineers, designers, municipal administration, cement industries involved in co-processing, cement manufacturers Association, other related stakeholders, waste management agencies, AFR traders, pre-processing agencies, and as a whole in the civil society. The book will also be helpful for the waste suppliers for cement plants involved in Co-processing. The book will be helpful for the researchers, students who work on resource efficiency, circular economy, waste management, policy making, environment management and engineering and in allied areas of academic fields.

This book will also serve as a reference book/course book to the students worldwide studying Environmental Science, Environmental Engineering, Civil Engineering, Chemical Engineering and Sustainability to learn the concept and business associated with cement kiln co-processing for sustainable management of waste materials. The book will be a treasure for the libraries.

We are hopeful that the book will be helpful for the readers and enhance the sustainability in waste management sectors through co-processing of AFRs.

Kolkata, India   Sadhan Kumar Ghosh
Mumbai, India   Ulhas V. Parlikar
Trondheim, Norway   Kåre Helge Karstensen
2021

Acknowledgements

We express our gratitude to the following organizations and individuals who supported during the preparation of the book.

Cement Manufacturers Association (CMA)
Centre for Sustainable Development & Resource
Efficiency Management, Jadavpur University
CII (GBC),
Consortium of Researchers in International
Cooperation (CRIC),
CPCB, New Delhi
Geocycle India,
GIZ,
Prof. S. P. Gautam, Ex-Chairman CPCB,
Dr. B. Sengupta and J. S. Kamyotra,
Ex-Member Secretary, CPCB.
Hardik Shah, IAS, Ex-MS, GPCB,
V. J. Anantharaman, Ex Director, D. I. Ltd.,
Monika Srivastava (Ms.), Srinivas Raju, ACC Ltd. K Ananth, K. Muralikrishnan, V. Kannan, CII(GBC),
Ravi Chikatmala, JSW
International Society of Waste Management, Air and Water (ISWMAW)
ISWMAW-IconSWM
Holcim,
Jadavpur University
M/s. Springer India Pvt. Ltd
Quality Management Consultants
The Royal Norwegian Embassy, New Delhi
SINTEF, Norway
TERI, New Delhi
Moumita Chakraborty (Ms.), Jean Pierre Degre, Ramesh Suri, Berthold Kren, Axel Peters and Deepak Ahuja, of Geocycle Ltd.

Milind Murumkar, Ex-Bharati Vicat Ltd.
Dr. S. B. Hegde, Pennsylvania State University.
Dr. Sannidhya Kumar Ghosh, UCB, USA.
Dr. J. D. Bapat, Consultant.
K. N. Rao, Consultant.

Cement industries and Equipment and System Suppliers for sharing details of their growth journey in co-processing through case studies, details description of equipment and systems for inclusion in this book for better understanding by the readers.

ACC Limited, India,
Ambuja Cement Ltd, India
ATS, France,
BEIL, India,
Dalmia Cement Ltd., India
GEPIL, India,
HIRAOKA, Japan,
J K Cement Ltd., India,
JSW Cement Ltd., India,
J K Lakshmi Cement Ltd., India
KHD Germany,
Loesche, Germany,
My home Industries Ltd., India Schenck Process, UK,
Sanghavi Engineering Pvt. Ltd., India
VICAT India,
VIMTA, India,

All researchers, friends, and others who helped directly or indirectly in preparing this book.

Sadhan Kumar Ghosh

Contents

Part III   Cement: Theory, Production Technology and Operations

4   Cement Manufacturing—Technology, Practice, and Development

About the Authors

Prof. Dr. Sadhan Kumar Ghosh having all degrees up to Ph.D. (Engg.) from Jadavpur University, is Professor in Mechanical Engineering and Chief Coordinator & Founder of, Centre for Sustainable Development and Resource Efficiency Management at Jadavpur University, India. He served as Dean, Faculty of Engineering and Technology at JU, and Director, CBWE, Ministry of Labour and Employment, Government of India, and Larsen & Toubro Ltd. He is a renowned personality in the fields of waste management, circular economy, green manufacturing, supply chain management, sustainable development, co-processing of hazardous and MSW in cement kiln, plastic waste and E-waste management and recycling, management system standards (ISO), and TQM having three patents approved. He is Founder Chairman of the IconSWM; Founder and President, International Society of Waste Management, Air and Water (ISWMAW); and Chairman, Consortium of Researchers in International Collaboration (CRIC). He received several awards in India and abroad including the distinguished visiting fellowship by the Royal Academy of Engineering, UK. He wrote 9 books, 52 edited volumes, and more than 230 national and international articles and chapters and is Associate Editor of IJMC&WM. His significant contribution has been able to place the name of Jadavpur University in the world map of research on waste management. He is consultant and international expert of UNCRD/DESA; Asian Productivity Organization; China Productivity Council; SACEP, Sri Lanka; IGES, Japan; SINTEF, Norway, etc. His research funding as PI includes EU Horizon 2020,

Erasmus +, UKIERI, DST, DBT, GCRF UK, Royal Academy of Engineering, Georgia Government, Jute Technology Mission, and many other funding agencies. He is involved in policy making bodies in India and standard development in ISO and BIS.

Mr. Ulhas V. Parlikar is Chemical Engineer with B. Tech from Osmania University and M. Tech from IIT Madras. He has also undergone the Senior Management Training Program (2010) and Senior Leadership Training Program (2012) from IMD, Switzerland. He has a total of 36 years of experience in Hindustan Lever Ltd, National Peroxide Ltd, & ACC Limited and has handled different responsibilities such as R&D, Process Engineering, Project Management, Business Development, Technology Collaborations, apart from building a large Geocycle (AFR) Business organization in ACC. He superannuated as Director of geocycle business at ACC Ltd. and Deputy Head of Geocycle India. Currently, he is extending consultancy services globally in the area of waste management, circular economy, policy advocacy, alternative fuels and raw materials (AFR), and co-processing. He successfully promoted the concept of co-processing in the Indian cement industry in the capacity of being Chairman of this initiative of Confederation of Indian Industries (CII) and successfully led the policy advocacy drive for the inclusion of co-processing as a preferred option for waste management in Indian policy framework. He has served as Member of many strategic committees of Holcim, ACC, and Geocycle and also as an expert on committees constituted by CPCB and Government of India. He has published >30 papers in national and international journals of repute with scientists, academicians, and officers of CPCB and SPCBs as co-authors. He has also served as Expert Member in drafting the Technology Roadmap Low-Carbon Technology for the Indian Cement Industry that was published by IEA and WBCSD.

Dr. Kåre Helge Karstensen received his Bachelor and Master of Science from the University of Oslo, Faculty of Natural and Mathematical Sciences, Department of Chemistry, and his Doctor of Philosophy from the Norwegian University of Science and Technology (NTNU), Faculty of Natural Sciences and Technology, Department of Chemistry. He is Chief Scientist at the Foundation of Scientific and Industrial Research (SINTEF) in Norway. SINTEF is one of the leading research organizations in Europe with more than 2000 employees from 75 countries. He has been working globally with circular economy, resource optimization, waste management, emission reduction, and co-processing for more than thirty years and is regarded to be a pioneer in the use of high-temperature cement kilns for management of hazardous wastes. He has tested and documented the feasibility of using cement kilns for environmentally sound destruction of hazardous organic chemicals like toxic pesticides, various persistent organic pollutants like DDT and PCBs, as well as ozone-depleting CFC gases. He has been instrumental in developing regulatory frameworks and guidelines for numerous governments, industries, and international organizations. He has been assisting leading international industry in raising their performance and sustainability benchmark. He has extensive experience from Asia and initiated and implemented co-processing practice in several Asian countries. He is Adjunct Faculty at Asian Institute of Technology in the period 2006–2016, where he is still active in research. He has published crucial R&D findings on the possibilities and limitations of co-processing in key scientific and technical journals with more than 120 peer-reviewed publications, books and chapters, >400 scientific reports, and >300 oral presentations at international conferences. He founded the company Global Sustainability in 2001 and is currently President and CEO.

Abbreviations

A/S	Alkali/sulphur
ACC	The Associated Cement Ltd
AF	Alternative Fuel
AFR	Alternative Fuels and Raw materials
AFRs	Alternative Fuels and Raw materials
AM	Alumina Modulus
APCD	Air Pollution Control Device
AR	Alternative Raw material
ATMA	India Tyre Manufacturing Association
AVL	Approved Vendor List
BAT	Best Available Techniques
BAT	Best Available Technology
BATs	Best Available Techniques
BEP	Best Environmental Practice
BIS	Bureau of Indian Standards
C&D	Construction and demolition
CCS	Carbon Capture and Storage
CCU	Carbon capture and utilization
CEMS	Continuous Emission Monitoring System
CII	Confederation of Indian Industries
CII	Confederation of India Industry
CISWI	commercial and industrial solid waste incineration units
CMA	Cement Manufacturers Association
CPCB	Central Pollution Control Board
CSI	Cement Sustainability Initiative
CSR	Corporate Social Responsibility
DE	Destruction Efficiency
DGFASLI	Directorate General Factory Advice Service & Labour Institutes
DMP	Disaster Management Plan
DRE	Destruction and removal efficiency
EIA	Environmental Impact Assessments

EJ	Exa Joules
EPR	Extended Producer Responsibility
ESM	Environmentally Sound Management
ESP	Electrostatic precipitator
EuCIA	European Composites Industry Associate
FAKS	Fluidized bed advanced cement kiln system
GBFS	Granulated Blast Furnace Slag
GCCA	Global Cement and Concrete Association
GCV	Gross Calorific Value
GEPIL	GreenGene Enviro Protection & Infrastructure Ltd
GFRP	Glass Fibre Reinforced Polymer
GHG	Green House Gas
GIZ	Deutsche Gesellschaft für Internationale Zusammenarbeit
GNR	Getting Numbers Right
GOI	Government of India
GTZ	Deutsche Gesellschaft für Internationale Zusammenarbeit
HHV	Higher Heating Value
HM	Heavy metal
HM	Hydraulic Modulus
HOWM	Hazardous and Other Wastes Management
IconSWM-CE	International Conference on Sustainable Waste Management and Circular Economy
ICPE	Indian Centre for Plastics in Environment
IEA	International Energy Agency
IFC	International Finance Corporation
ILC	In Line Calciner
INDC	Intended Nationally Determined Contribution
IPCC	Intergovernmental Panel on Climate Change
IPPC	Integrated Pollution Prevention and Control
IPPC	International Plant Protection Convention
ISWMAW	International Society of Waste Management, Air and Water
IUCNNR	International Union for Conservation of Nature and Natural resources
LCA	Life Cycle Analysis
LCTR	Low-Carbon Technology Road
LHV	Lower Heating Value
LOI	Loss on ignition
LSF	Lime Saturation Factor
MNRE	Ministry of New and Renewable Energy
MoEF	Ministry of Environment and Forest
MoEFCC	Ministry of Environment, Forest and Climate Change
MoHUA	Ministry of Housing and Urban Affairs
MoNRE	Ministry of Non-renewable Energy
MSDS	Material Safety Data Sheets
MSW	Municipal Solid waste

MTPA	Million tons per annum
NAAQS	National Ambient Air Quality Standards
NCV	Net Calorific Value
NCV	Gross Calorific Value
NIMBY	Not In My Back Yard
NRPW	Non-Recyclable Plastic waste
NSR/PSD	New Source Review/Prevention of Significant Deterioration
ODS	Ozone Depleting Substances
OH&S	Occupational Health and Safety
OH&S	Occupational Health and Safety
OPC	Ordinary Portland Cement
PCC	Portland Composite Cement
PCDD	Polychlorinated dibenzo-p-dioxin
PCDD/Fs	polychlorinated dibenzo-p-dioxins and dibenzofurans
PCDF	Polychlorinated dibenzofuran
PICs	Products of incomplete combustion
PLC	Portland Limestone Cement
PM	Particulate matter
POHC	Principal Organic Hazardous Compounds
POPs	Persistent Organic Pollutants
POPs	Persistent Organic Pollutants POPs
PPC	Plain cement concrete
PPC	Portland Pozzolana Cement
PPE	Personal protective equipment
PSC	Portland Slag Cement
PWM	Plastic Waste Management
QAP	Quality Assurance Plans
RDF	Refuse-Derived Fuel
RDF	Refuse-Derived Fuel
RII	Resource Intensive Industries
RIIs	Resource Intensive Industries
RPM	Rounds-per-minute
SBM	Swachh Bharat Mission
SCF	Segregated Combustible Fraction
SCR	Selective Catalytic Reduction
SDGs	Sustainable Development Goals
SIR	Sustainability Index of Recovery
SLC	Separate Line Calciner
SLF	Storage Land filling
SM	Silica Modulus
SNCR	Selective Non-Catalytic Reduction
SOP	Standard Operating Procedure
SOPs	Standard Operating Procedures
SPCB	State Pollution Control Board
SPL	Spent Pot Liner

SSEF	Shakti Sustainable Energy Foundation
SUPs	Single Use Plastics
SWM	Solid Waste Management
TDF	Tyre Derived Fuel
TOC	Toxic Organic Compounds
TOE	Ton of oil equivalent
TPD	Tons per day
TSDFs	Treatment, Storage and Disposal Facilities
TSR	Thermal Substitution Rate
UNDP	United Nation Development Programme
UNEP	United Nations Environment Programme
VOC	Volatile Organic Compound
VSK	Vertical Shaft Kiln
WBCSD	World Business Council for Sustainable Development
WFD	Waste Framework Directive
WTE	waste-to-energy
WWF	World Wide Fund for Nature
XRD	X-ray Diffraction

List of Figures

List of Tables

Part I
Introducing the Subject

Chapter 1
Introduction

1.1 Backdrop

The cement manufacturing is an energy-intensive process which is producing cement, the sought for materials for building construction around the world. It generally consumed in the order of 3.3 GJ/ton of thermal energy of clinker produced and nearly 90–120 kWh of electrical energy/ton of cement is produced (Giddings et al., 2000; European Commission, 2001). Indian Cement industry has the best performance numbers and the best among the Indian industry is 2.83 GJ thermal energy per Ton of clinker and best electrical energy consumption is 56.14 KWH/T cement (CII, 2021).

Traditionally, cement plants use coal as the primary fuel. A range of other fuels, such as oil, gas, petroleum coke different types of liquid wastes and solid waste, via co-processing have also been successfully used as energy sources for firing in kilns for cement production. However, Natural Gas is only used in a few places, e. g., Middle East, and usually as primary fuel. Cement production is responsible for nearly 2.4% of global CO_2 emissions from industrial and energy sources during 1987–88 (Marland et al., 1989) while during 2019–20 it contributes to nearly 7.0%. Cement production is the third largest consumer of energy @ 9–10% of the total industrial energy consumed and CO_2 emission ranges from 0.7 to 0.93 ton/ton of cement produced. As estimated in 2018, 1445 million tons/year of CO_2 was emitted by the top ten cement producing countries including the European Union with the strategy of incorporating low carbon cement manufacturing process, such as reducing clinker factor, improving energy efficiency, and co-processing of wastes as AFRs leading to carbon mitigation. As clinker production emits CO_2, the IPCC recommends to use clinker data instead of taking the cement data to estimate CO_2 emissions. Cement could potentially be imported from the producer country and hence, in the international trading of clinker the cement production data may create biased emission estimates. The amount of clinker produced in blended and natural types of cement is highly variable and the estimation depends on the clinker factor.

© The Author(s), under exclusive license to Springer Nature Singapore Pte Ltd. 2022
S. K. Ghosh et al., *Sustainable Management of Wastes Through Co-processing*,
https://doi.org/10.1007/978-981-16-6073-3_1

Since the beginning of the 1970s wastes materials from different sources have successfully been co-processed as AFR in cement kilns in a few countries, namely Australia, Canada, Europe, Japan, and USA. However, everything started in the US in 1972 and subsequently the process was implemented in other countries. Incineration process is widely used worldwide to reduce the burden of municipal wastes, hazardous wastes and hazardous chemical, biological waste, reduction of potential toxicity and in very rare cases ton reduce the potential infectious properties and volume of medical waste. In 80 and 90 s, there were confusion over the means of appropriately managing waste. Pollutants like polychlorinated dibenzo-p-dioxins (PCDDs), and polychlorinated dibenzofurans (PCDFs) emissions were discovered and identified to be coming out from MSW-incinerators, etc., in the decade 1980–1990 which is major concern as the potential risk to human health and the environment at certain higher concentrations than the prescribed levels, however had nothing to do with co-processing of wastes in cement kilns.

In the cement production, raw material is fed into the rotary kiln. Appropriate types of hazardous wastes are used as auxiliary or replacement fuel that are burnt under controlled atmosphere. Since the year 1972 hazardous waste are burnt in cement kilns, which was practiced in Ontario, Canada, United States, Belgium, and Switzerland started using waste-fuelled kilns. The use of wastes as fuel in the kilns is the energy recovery initiative that led to the conservation of non-renewable fossil fuels and circulation of wastes enhancing resource efficiency associated with an economic incentive paid to the kiln operators. Hazardous wastes were burnt in light-weight aggregate kilns and different types of furnaces as well in cement kilns. In last three decades, the situation in respect of use of waste materials as Alternative Fuels is totally changed. Earlier, the cement kilns used to get paid for co-processing wastes as Alternative Fuels in cement kilns. However, at present, due to popularity of co-processing of wastes in cement plants, in many of the locations the business model has got revised where wastes are to be purchased by the cement plant operators from waste generators.

The use of waste as AFR in resource and energy-intensive industries is called co-processing. Co-processing in cement kilns is a technology practiced worldwide as a sustainable solution to the management of wastes from municipal sources, agricultural sources, and industrial sources. AFR from waste definitely contribute reducing the cost as well as the amount of fossil fuel use helping in conservation of natural resources, reducing the landfills possibilities, and curbing the global CO_2 and GHG emissions.

Significant advancement has taken place in the scientific and technological aspects of co-processing while developing and implementing this technology in the cement plants has taken a faster pace in operational scales during 2010 to the present in a few developing countries. This technology is in practice for more than three decades and has been recommended by Basel Convention for the sustainable management of Hazardous wastes and by the Stockholm Convention for the sustainable management of POPs. Using co-processing technology is a part of the waste management rules notified by the ministry of the government of India in 2016 and has been provided as a sustainable waste management option over the conventional options of incineration and landfill. Co-processing promotes curbing the climate change impacts and global warming, conserving natural resources and to implement 3R and circular economy concepts on large scale.

Co-processing like other waste management initiatives has potential to provide large scale employment. Co-processing provides a sustainable solution for industrial wastes if used in cement kiln because the waste is used as fuel as well as the raw materials leaving no residue to be disposed to landfill. This is one of the advantages of co-processing compared to MSWI, which produce 20–30% ash, slag, flue-gas cleaning products, etc. The fossil fuels are substituted by combustible parts of the waste and the raw materials, such as Silica and Iron are replaced by non-combustible parts of the waste (Ash) which in turn acts as substitute raw material. The instant use of energy released without any losses results in the highest value of energy efficiency in cement kiln. The adverse impact on the environmental is negligible. The cement industries in several countries are encouraged to realize the potential of use of wastes as alternative fuels and raw material (AFR) in cement kilns with several benefits. Indian cement industries have taken a target to reach a 25% thermal substitution rate (TSR) by 2025, from the current level of less than two-digit percentage. It has been estimated that to achieve 25% TSR at 2025, Indian Cement Industry will require 7.07 million TOE of energy from Alternate fuels. While "TOE" is expanded as ton of oil equivalent (TOE), a unit of energy. The amount of released energy by burning one ton (1000 kg) of crude oil is called TOE.

Even though this technology has received the required attention and inclusion in the policy framework of government of India, its understanding and awareness with the stakeholders belonging to the academic and other relevant sections is vastly missing. This is because of lack of appropriately researched and collated information on this topic particularly in the context of Indian wastes and its management. The students of the environmental faculty continue to get focused education on the incineration and landfill options and not yet much about the co-processing technology. The understanding of co-processing technology is still very limited to the involved stakeholders which needs mass awareness in the countries. The number of waste-to-energy and incineration plants are quite low in the globe in comparison to the number of cement kilns involved in co-processing of wastes; while the awareness regarding waste-to-energy and incineration plants is more than co-processing of wastes among the policy makers, scientists, and general citizens. As there is a huge potential of co-processing of waste that exists worldwide as a sustainable waste management option, a concerted drive for enhancing awareness on co-processing is required to be established. Increasingly effective co-processing techniques in cement kiln have been developed in several developed countries, in India and China over the last 40 years. More than 60 factories in India and 200 factories in China are involved in co-processing of municipal solid waste (MSW) derived and other types of wastes.

1.1.1 Sustainable Development and Waste Management

The most recent and popular initiative of the United Nations to leverage sustainable development in the globe until 2030 is the Sustainable Development Goals. The SDGs are seen in the report "Transforming Our World", predicted at "The Future We Want". The publication of the "World Conservation Strategy" by the United Nations Environment Programme (UNEP), World Wide Fund for Nature (WWF), and International Union for Conservation of Nature and Natural Resources (IUCNNR) was considered to be the milestones initiative for sustainable development. The strategy adopted in these was the precursors to the concept of sustainable development, which aimed:

- Maintaining fundamental ecological processes and life-support systems, vital for human survival and development.
- Preservation of genetic diversity, which depend on the breeding programmes needed for the protection of plants and domesticated animals, as well as much scientific innovation, and the security of the many industries that use living resources.
- Ensuring the sustainable utilization of species and ecosystems (notably fish and other wildlife, forests, and grazing lands), which supports millions of rural communities as well as significant industries.

In 1987, the Brundtland Report "Our Common Future" defined "sustainable development" as: "The development that meets the needs of the present without compromising the ability of future generations to meet their own needs."

Two key concepts are emphasized in the Brundtland Report (WCED, 1987) as excerpted below

- _ "needs," in particular the essential needs of the world's poor, to which overriding priority should be given, and
- _ "limitations" imposed by the state of technology and social organization on the environment's ability to meet present and future needs.

In comparison, sustainable development was defined by the President's Council on Sustainable Development in the United States as (USEPA, 2013): "…an evolving process that improves the economy, the environment, and society for the benefit of current and future generations."

"Agenda 21: A Programme of Action for Sustainable Development," establishes 27 principles around the 3 pillars of sustainability: economy, society, and environment, which is known as The Rio Declaration on Environment and Development, was adopted in 1992. Stated that the critical dimensions of sustainable development should be "The moral imperatives of satisfying needs, ensuring equity and respecting environmental limits. The model reflects both moral imperatives laid out in philosophical texts on needs and equity, and recent scientific insights on environmental limits."

The intention of this book goes much further in the establishment of sustainable Waste management through a technology where several other technologies of waste management are available which in turn helps in achieving the Sustainable Development Goalsss.

According to the: "Waste means any substance or object which the holder discards or intends or is required to discard." The wastes are classified by source, nature, physical and mechanical properties, chemical and elemental properties, biological/biodegradable properties, and combustion properties. The wastes categories are defined in the rules of respective countries, such as Solid Waste Management Rules 2016 defines the categories of wastes in India. Waste disposal is an important issue in the world. Each of the types of Wastes has a value chain which has not been tapped in many areas and in several countries too. Waste is considered as very important entity considered as resource to be utilized using different technologies. The utilization of these resources which would have been wasted can help in reducing extraction of natural resources leading to sustainable development. This book deals with one of such technologies, called the co-processing technology where waste is utilized in cement kiln with 100% recovery potential. The sustainability of different waste management technologies has been assessed in Chap. 3 of this book to understand the sustainability status of co-processing of waste in cement kiln.

Landfill and incineration are the easiest and popular disposal practices in many countries (Corinaldesi et al., 2015; Ribeiro et al., 2015). To be specific, in Ireland, there is no law in banning the disposal of turbine blade waste into landfill which exists in Germany (Correia et al., 2011). Landfilling costs in Ireland is 113 Euro per ton, one of the most expensive rates in Europe (National Competitiveness Council, 2015). Glass fibre reinforced polymer (GFRP) is the main component of wind turbine blades. The cost of virgin materials to produce GFRP is very low and hence, pyrolysis and mechanical processing of GFRP is not viable according to the European Composites Industry Associate (EuCIA). The co-processing of GFRP in a cement kiln is being practiced as a viable option. The cost of co-processing is still unknown, however, is estimated to be more than landfilling. In compliance to the rules, co-processing is a sustainable and viable option. Similar way, the co-processing is a sustainable solution to the waste management.

1.2 Co-processing in 70 and 80s

In the late 70 and 80 s, the burning of wastes in cement kilns had lot of issues with respect to the emission. The combustion efficiency of CO and total hydrocarbons is a main characteristic to be analysed in a cement plant. Researchers could not establish relationship between good combustion practice and emission of dioxin. The possible reasons are (a) total hydrocarbons and CO are associated with the raw-mineral feedstock, rather than the fuel and (b) because of high combustion temperatures in the kiln, nonequilibrium conditions may evolve the CO. Injection of powdered activated carbon is adopted as practiced many times for the removal of dioxins, furans, and mercury in municipal-waste and hazardous-waste incinerators.

Co-processing is an effective waste utilization option than discarding to the land-fills of incineration. Wastes with high calorific value are used as AFR to replace fossil fuels in the cement kilns. There must be special control of byproducts of chlorine and hydrocarbon, such as polychlorinated dibenzo dioxins (PCDDs) and polychlori-nated dibenzofurans (PCDFs) in incineration processes. The cooling process in "De novo synthesis" from 450 to 200 °C help in forming dioxins and furans. A trial burn is needed to measure the unintentional by products that may be helpful for under-standing phenomenon that takes place in the process in the kiln. In a few countries, the trial burn has been carried out while it took place in the Bulacan cement plant of the Union Cement Corporation in the Philippines way back in November 2004.

1.3   Co-processing in 90s Until 2010

More interest in co-processing has been observed and demonstrated by the researchers, cement manufacturers and manufacturer's association and government. Co-processing of wastes helps to achieve the targets in Agenda 21 in "Earth Summit" in, the Johannesburg Declaration on Sustainable Development (2002), and the Sustainable Development Goals 2030.

The GTZ-Holcim Public Private Partnership developed the guidelines for Co-processing in 2006 for the respective stakeholders and decision makers. A few guiding principles presented in Table 1.1 gives a general condition where co-processing can be applied. The excerpts of the guiding principles have been presented in the Table 1.1. The Guidelines took reasonable consideration on international conventions, namely the Basel and the Stockholm Conventions and the UN Framework Convention on Climate Change (Kyoto Protocol). The guidelines proposed twenty-two principles broadly covering legal aspects, impact on Environment, operational issues, aspects related to communication and social responsibility and occupational health and safety (OH&S) issues.

The guidelines proposed the general principles for co-processing demonstrated in Table 1.1.

1.4   Co-processing in Post 2011

India is the second largest producer of cement next to China. It has improved the energy efficiency of cement plants in recent past. In 2012, the TSR, of the plants by using alternative fuels, e.g., biomass and municipal waste, stands at only 0.6% while it reaches more than 4% in 2020. This was made possible by partial substation of fossil fuels with considerable efforts by all stakeholders. Efficient waste collection systems and pre-processing facilities are to be made robust, while the industry needs

Table 1.1 Proposed general principles for co-processing in cement plant. *Source* Guidelines on co-processing Waste Materials in Cement Production, 2006

Principle number	Statement of principles	Expansion of the statement of principles
I	"Co-processing respects the waste hierarchy"	Co-processing supports 3Rs and the waste hierarchy starting from waste reduction and least priority to landfilling. Co-processing is eco-friendly resource recovery option in waste management and comply with the Basel and Stockholm Conventions
II	"Additional emissions and negative impacts on human health must be avoided"	Zero or minimum adverse impact on the environment and risks to human health which should be less than that from cement production using fossil fuel
III	"The quality of the cement product remains unchanged"	Heavy metals from the product (clinker, cement, concrete) shall be removed and adverse impact on the environment should be zero or minimum as, for example, leaching test demonstration. The cement thus produced will allow end-of-life recovery
IV	"Companies engaged in co-processing must be qualified"	The company that is involved in co-processing must have good track records of compliance to environmental and safety and a high level of commitment to protect the environment, health, and safety. Relevant and right information to be shared with the public and the authorities The company should demonstrate due diligence, commitments of compliance to applicable rules and regulations. It control parameters in the processes and sub-processes to achieve efficient co-processing, relationship with the public and other appropriate actors in all spheres of local, national, and international arena

(continued)

Table 1.1 (continued)

Principle number	Statement of principles	Expansion of the statement of principles
V	"Implementation of co-processing has to consider national circumstances"	Regulations and procedures should address need and requirements and needs specific to the national boundary. A long-term implementation plan is required to build required capacity and institutionalizing the arrangements

to continuously invest in co-processing and alternative fuel feeding systems with visible support of the government in the provision of incentives for the processing of municipal wastes. Cement industries have set a goal of achieving 25% TSR by 2030, which is an ambitious target that will bring significant improvement in fossil fuel substitution in cement industries reducing huge burden of waste management.

In the decade 2011 to 2020, significant improvement in the areas of co-processing in cement kilns has taken place. A number of guidelines documents (Basel Convention, 2012; CPCB, 2017; CPHEEO, 2018; LafargeHolcim-GIZ, 2020) were developed and published containing the guidelines on Pre-processing and Co-processing of Waste in Cement Production—Use of waste as alternative fuel and raw material. A revised edition of the 2006 guidelines was published in 2020 to address to the stakeholders and decision makers to update technical, institutional, legal and social aspects of the original document to support continuous improvement in the application of pre-processing of waste in the cement industry. The first edition of the guidelines was focused on co-processing of industrial and commercial waste (GIZ-Holcim, 2006), while the revised guidelines was published with a stronger emphasis on pre-processing of wastes into alternative fuel and raw materials (AFR), pre- and co-processing of municipal waste and integrating pre- and co-processing into local waste management value chains. More information is given on how pre- and co-processing contributes to the sustainable development goals, its climate relevance, financing and ways to work with the informal waste sector. The CPHEEO, under the ministry of Housing and Urban Affairs (MoHUA) in India, constituted an Expert Committee in October 2017 to prepare "Norms for Refuse Derived Fuel (RDF) from Municipal Solid Waste for its utilization in cement kilns, waste-to-energy plants and similar other installations" to achieve the objectives of a clean India under Swachh Bharat Mission (SBM). The intended objectives of preparing the document were to enhance the use of MSW-based RDF of significant quality of desired standard in industries to comply with the SWM Rules, 2016. The Committee came up with the Guidelines on Usage of RDF in Various Industries in 2018 for co-processing in cement plants and incineration plants, etc.

In last couple of years, since the year 2014, due to three major initiatives by the government of India, the waste segregation, collection, and pre-processing facilities have been strengthened. The initiatives are (1) Swachh Bharat Mission (SBM) launched in 2014, (2) Revision and notification of six rules of waste management in the year 2016, and (3) introduction of Swachh Survekshan in 2016.

1.5 National Rules, Regulations, Standards Supporting Co-processing

National rules, regulations, standards, and the technical infrastructure in the country is very important prime mover for implementation of co-processing of different types of wastes. The developing countries are less mature than in countries that have a long experience with co-processing waste in the cement industry. The directives for co-processing in European countries are matured and have been in implementation for quite a long time. The Solid Waste Management Rules 2016 and Hazardous and Other Waste Management Rules 2016 in India have specifically mentioned co-processing of waste in cement kiln as an option of waste treatment.

Many developing countries such as India, China, and Southeast Asian countries are initiating programmes to promote co-processing of wastes in the cement industry. Effective regulatory and institutional frameworks are critical to ensuring that co-processing practices in the cement industry are not harmful to health or the environment. An integrated solid waste management model and regulations and standards related to environmental performance, product quality, operations and safety, permitting procedures, monitoring and reporting are key factors in a regulatory framework and policy instrument for a sustainable co-processing industry. From the technological perspective, pre-processing and treatment of waste are often required to make the waste ready for co-processing in cement kilns. As much as possible, best available techniques (BATs) should be applied to the pre- and co-processing processes in order to ensure that waste co-processing in the cement industry is environmentally sound. Chapter eight of the book discussed the details of Waste Management rules focusing on co-processing in different countries.

1.6 Criteria for the Success of an Alternative Fuel Project

The success of project involving co-processing and AFR depends on several factors. A few of the key factors may include

- Management commitment.
- Plant capability to respond to regulatory changes.
- unfair competition against co-processing and lobbying to support regulatory barriers against co-processing.
- levels of investment.
- Downstream of the regulatory framework, permitting the use of waste as AFR is often a complex process and takes lot more time.

- Establishing trustworthy relationships with stakeholders, particularly local residents is crucial for the success. The local residents are important stakeholder to have awareness on need & implications (regarding pollution, health, and safety) of waste treatment in a cement plant located in the vicinity.
- good knowledge of the various waste supply chain, the market and existing competitive price in the country and locality.
- Understanding the kiln's ability to replace fossil fuels, technical aspects, necessary actions and process control parameters to achieve the TSR.
- Control parameters of waste pretreatment is critical to the quality. Uncontrolled liquid mixture may result in variations in calorific value. The operator of the facility must have in-depth knowledge of the constraints of the cement kilns.
- Pre-Processing effective transformation of inhomogeneous waste into homogenous AFR for co-processing is essential.
- Effective and transparent communication between the two actors- waste supplier and plant operator is key to the success of co-processing.

In recent past increased focus is given on formulating strategies of carbon mitigation in different energy-intensive manufacturing processes. Co-processing is found to be the most prominent low carbon cement manufacturing process.

Different definitions and terminologies have been introduced in the next chapter which will be helpful throughout the book .

1.7   The Structure of This Book

Chapter 1: Introduction

This chapter provides the overview on the content of the book, their relevance in the current context of sustainable development and circular economy. The audience of the book include students of environmental engineering, civil engineering, chemical engineering, sustainability, etc., cement plants and executives working therein, waste management plants and their executives working therein, professionals associated with the subject of co-processing and waste management, authorities, consultants, project management professionals, academicians, and researchers. The audience addressed is global with substantial focus on the developing world because it is here that the main requirement of such a book is

---

Chapter 2: Terms and definitions related to co-processing

This chapter provides the definitions of various terms related to the subjects of co-processing. These terms belong to the subject of cement, cement chemistry, waste materials, waste management, sustainability, thermal characteristics, safety, operation and management, quality monitoring and control, economics, etc.

(continued)

(continued)

Chapter 3: Status review of research on co-processing
This chapter provides the review of various research and applications initiatives related to the technology of pre-processing and co-processing of waste materials. The various aspects considered in the review are use of the waste materials in cement and allied industries as a resource; challenges faced in the applications and solutions implemented; policies and their impacts; impact the technology has made on the sustainable management of wastes, etc.
Chapter 4: Cement manufacturing—Technology, Practice, and Development
This chapter provides in detail the types of cement, various technologies associated with its manufacture, the operational aspects of these technologies, etc. The cement manufacture consists of various operations such as quarrying, preparation of raw materials, fuel preparation, clinker manufacturing process, cement grinding, and cement dispatch. There are different technologies utilized in the manufacture of cement such as dry process with and without pre-heater/pre-calciner arrangement, semi dry process, semi-wet process, wet process, and vertical shaft kiln. Depending upon the technology adopted the various operations starting from quarying to cement dispatch would vary and these are described in detail. The type of technology also has bearing on the way the co-processing of AFRs would be implemented and same is also addressed in this chapter
Chapter 5: Fundamentals of Cement Chemistry, Operations and Quality Control
Cement manufacture involves considerable science dealing with the chemistry with which the different raw materials react at high temperature to form the clinker as an intermediate product and then its mixing with gypsum and pozzolanic materials to manufacture the cement product. The different raw materials are consisting majorly of CaO, Fe_2O_3, Al_2O_3, and SiO_2 and the high temperature of reaction is attained using coal or other fossil fuels and AFRs in the rotary kiln. This being a chemistry driven process, there are several quality control steps involved through which the chemistry of the raw mix is ascertained in the kiln and desired quality product is manufactured. The process of cement manufacture also requires precise control over the process parameters to ensure optimum utilization of the resources and the cost. This chapter provides details in respect of this aspect of cement manufacture and co-processing implemented during the same
Chapter 6: Guidelines on co-processing of AFRs—International best practices
This chapter covers all the important aspects related to implementing co-processing in practice. These include required legal framework and policies, general requirements, requirements pertaining to the health and safety, operational considerations, quality monitoring and control, emission-related aspects, storage and handling, community concerns, external communications, labelling and traceability of the wastes, pre-processing of wastes into AFRs, etc.
Chapter 7: Sustainability considerations in cement manufacture and co-processing -
Cement production causes considerable impacts on the environment due release of GHG and other emissions, biodiversity due to quarrying, water utilization in process causing reduction in the water table, plastic utilization causing plastic waste management issue, reduction in the natural (fossil) resources, etc. Cement industry has therefore taken up several sustainability initiatives globally and the same is being monitored across the globe through appropriate agencies. Several challenges encountered while implementing the sustainability initiatives in the cement industry are also discussed. This chapter provides an overview on these different sustainability initiatives taken up by the cement industry globally—including co-processing of AFRs—and provides the summary of the success achieved with the same KPI

(continued)

(continued)

Chapter 8: Waste Management rules focusing on co-processing in different countries

Policy framework is an important aspect for ensuring safe and environment friendly management of wastes. Co-processing provides waste management in an environmentally sound manner. Different countries that have set up their own policy frameworks in the form of Acts, Rules, guidelines, etc., have promoted co-processing in the same for the management of wastes rather than sending them for incineration or landfilling. Basal convention has also brought out their own guidelines for management of hazardous wastes and POPs using co-processing. This chapter reviews the policy framework of some of the developed countries and under developed countries and provides the overview of the policy framework of India which is suitable for adaptation by the developing countries

Chapter 9: Emission considerations in cement kiln co-processing

For successful co-processing, monitoring and control of emission within prescribed limits is an important requirement. Co-processing is not supposed to influence the emissions from the cement kilns and the policy frameworks of all the countries have mandated values for compliance for the different streams. This chapter discusses the specified standards for emission control while undertaking co-processing and on-line monitoring of the same using continuous emission monitoring system

Chapter 10: Co-processing of wastes as AFRs in cement kilns

This chapter discusses the salient features of co-processing and their importance in waste management. It also reviews the co-processing feasibility of different kinds of wastes from municipal, Industrial, and agricultural sectors that can be co-processed. The banned wastes for pre-processing and co-processing are appropriately touched. The chapter also discusses the importance of pre-processing before co-processing and the infrastructure required to undertake co-processing. Various parameters that need to be monitored for efficient and successful co-processing including control of emissions are also discussed in this chapter. Co-processing is required to be undertaken only after obtaining required permissions and this is also discussed in the chapter. Apart from discussing the principles of co-processing, the chapter also discusses various technological aspects of the same starting from receipt of wastes to co-processing them by feeding them into the kiln. Different manual and mechanized schemes are discussed in detail

Chapter 11: Pre-processing of wastes into AFRs

This chapter explains the concept of pre-processing and provides detailed explanation on the various unit operations and equipment associated with pre-processing of wastes into AFRs. It also discusses the category-wise pre-processing schemes and infrastructure required for the same. Important aspects of quality control, health, and safety are also discussed in detailed in the book. Flow schemes for pre-processing of solid, liquid, and pasty wastes are explained in detail in this chapter

Chapter 12: Operational considerations in co-processing

Co-processing of AFR is a pretty complex process on account of different level of moisture, ash, and burnability characteristics compared to conventional materials. Therefore, several operational challenges are faced while undertaking co-processing. Higher level of Chlorine, Alkalies, and Sulphur present in AFRs bring additional complexity in the operation of the cement kiln. The operational considerations also pertain to man power requirement, legal aspects, process and project design, stakeholder engagement and external communications, material handling, quality parameters, etc. These various aspects are elaborated in the Chap. 12

(continued)

(continued)

Chapter 13: Case Studies and business models related to Cement Kiln co-processing

AFR co-processing needs to be implemented in a business mode by devising different business model. The commercial considerations towards AFRs vary depending upon their liability considerations. These need to be factored appropriately in the business model for desired success rate in the same. The business models could be manual or mechanized depending upon the nature of material. These business models are elaborated in detail in this chapter by taking typical case study data and example

---

Chapter 14: Global Status on co-processing

Co-processing is in practice for more than three decades now and is being implemented in several countries. This is one of the strong pillars for reducing the carbon foot print of the cement industry. Cement Sustainability Initiative is a common programme of a group of cement companies that represent about 30% of the cement capacity worldwide. CSI is dedicated to the cause of sustainability, and it is documenting the efforts being made by them in a database names as "Getting the Numbers Right (GNR)." This chapters utilizes the information presented in this database to understand the status of co-processing in developed country, in developing country and in the world. It also discusses the status of co-processing in some of the countries providing information in this database

---

Chapter 15: Journey of the Growth of Co-processing in India

Recognition of co-processing as a preferred option for the management of wastes in India has gone through a very detailed evolution phase. The learning from this evolution phase is very useful to other developing countries who have to implement co-processing in the cement kilns of their country. Before receiving desired recognition, several initiatives have been implemented in the country which involved demonstration of the technology through >90 successful co-processing trials, implementation of multi-stakeholder consultation processes, large scale awareness and capacity building initiatives, and considerable policy advocacy efforts. This chapter also illustrates the case studies of some of the cement companies from India—that represent more than 25% of the cement capacity of the country—in successfully traversing their AFR co-processing journey

---

References

BASEL CONVENTION. (2012). Technical guidelines on the environmentally sound co-processing of hazardous wastes in cement kilns, Basel Convention, UNEP, © 2012 Secretariat of the Basel Convention, Secretariat of the Basel Convention International Environment House 11–13 chemin des Anémones 1219 Châtelaine, Switzerl.

Correia, J. R., Almeida, N. M., & Figueira, J. R. (2011). Recycling of FRP composites: Reusing fine GFRP waste in concrete mixtures. The Journal of Cleaner Production, *19*, 1745e1753. https://doi.org/10.1016/J.JCLEPRO.2011.05.018.

Corinaldesi, V., Donnini, J., & Nardinocchi, A. (2015). Lightweight plasters containing plastic waste for sustainable and energy-efficient building. Construction and Building Materials. https://doi.org/10.1016/j.conbuildmat.2015.07.069.

CPCB. (2017). Guidelines for Co-processing of Plastic Waste in Cement Kilns (As per Rule '5(b)' of Plastic Waste Management Rules, 2016) by CENTRAL POLLUTION CONTROL BOARD Ministry of Environment, Forest and Climate Change, Government of India in May, 2017.

CPHEEO. (2018). Guidelines on Usage of Refuse Derived Fuel in Various Industries, prepared by Expert Committee Constituted by Ministry of Housing and Urban Affairs (MoHUA), Central

Public Health And Environmental Engineering Organisation (CPHEEO), Ministry of Housing and Urban Affairs, Govt of India.

CII, Energy Bench-marking of the Indian Cement Industry, Ver. 5.0 (2021, May).

European Commission (EC). (2001). Integrated Pollution Prevention and Control. Reference Document on Best Available Techniques in the Cement and Lime Manufacturing Industries.

Giddings, D., Eastwick, C. N., Pickering, S. J., & Simmons, K. (2000). Computational fluid dynamics applied to a cement precalciner. Proc. Instn. Mech. Engrs. Vol. 214 Part A.

GIZ-Holcim. (2006). Guidelines on co-processing Waste Materials in Cement Production, The GTZ-Holcim Public Private Partnership, Copyright © 2006 Holcim Group Support Ltd and Deutsche Gesellschaft für Technische Zusammenarbeit (GTZ) GmbH, Federal Ministry of Economic Cooperation and Development, Germany.

Intergovernmental Panel on Climate Change (IPCC). (1997). Revised 1996 IPCC Guidelines for National Greenhouse Gas Inventories. Reference Manual (Revised). Vol 3. J.T. Houghton et al., IPCC/OECD/IEA, Paris, France.

LafargeHolcim-GIZ. (2020). Guidelines on Pre- and Co-processing of Waste in Cement Production-Use of waste as alternative fuel and raw material, Copyright c 2020 Holcim Technology Ltd and Published by Deutsche Gesellschaft fur Internationale Zusammenarbeit (GIZ) GmbH.

Marland, G., Boden, T. A., Griffin, R. C., Huang, S. F., Kanciruk, P., & Nelson, T. R. (1989). Estimates of CO_2 Emissions from Fossil Fuel Burning and Cement Manufacturing, Based on the United Nationals Energy Statistics and the U.S. Bureau of Mines Cement Manufacturing Data. Report No. #ORNL/CDIAC-25, Carbon Dioxide Information Analysis Centre, Oak Ridge National Laboratory, Oak Ridge, Tennessee, USA.

Ribeiro, M. C. S., Meira-Castro, A. C., Silva, F. G., Santos, J., Meixedo, J. P., Fiúza, A., Dinis, M. L., & Alvim, M. R. (2015). Re-use assessment of thermoset composite wastes as aggregate and filler replacement for concrete-polymer composite materials: A case study regarding GFRP pultrusion wastes. *Resources, Conservation and Recycling.* https://doi.org/10.1016/j.resconrec.2013.10.001.

Chapter 2
Terms and Definition Related to Co-processing of Waste in Cement Kiln

2.1 Introduction

The definition of a term is a word or group of words that has a special meaning, a specific time period or a condition of a contract. In other words, a definition is a statement of the meaning of a term. The term must be explained in specific way to understand the meaning of the term. There may be conceptual and operational definitions. The conceptual definition is abstract and most general in nature. The usual source of conceptual definition is the dictionary which is the reference book for everyday language. Operational definition of terms or the concept refers to a detailed explanation and meaning used in a particular study. Unlike the conceptual definition, it states in concrete term in that it allows measurements for data collection, where necessary. The usual practice when using both types of definition is to state first the conceptual followed by the operational definition. Several guidelines, reports, national standards, and research documents evolved different terms and definitions related to co-processing. This chapter is a compilation of all such terms and definitions. However, there may be some terms and definitions which are available in the respective national legislation and regulations in respective countries that may differ a little from the definitions in this chapter.

There are many terms related to co-processing and associated fields which have been used in this book. It is necessary that these terms must be defined beforehand to make an easy understanding while reading different chapters in the book. The terms and definitions have been classified in four categories in this book for ease in understanding: (a) Materials-related terms and definitions, (b) Process and system-related terms and definitions, (c) Legal and permit-related terms and definitions, (d) Infrastructure-related terms and definitions, and (e) Stakeholder-related Terms and Definitions. Following sections will define different terms which is depicted in Fig. 2.1.

© The Author(s), under exclusive license to Springer Nature Singapore Pte Ltd. 2022   17
S. K. Ghosh et al., *Sustainable Management of Wastes Through Co-processing*,
https://doi.org/10.1007/978-981-16-6073-3_2

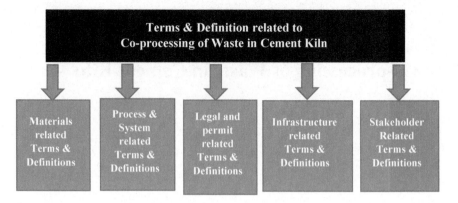

Fig. 2.1 Classification of terms and definition related to co-processing

2.2   Materials-Related Terms and Definitions

Different types of materials are related to co-processing and associated activities. The materials-related terms which are associated directly with the co-processing are Alternative fuels and raw materials (AFR), refuse derived fuel (RDF), etc., while Cement, clinkers, Concrete, etc., are associated with the demand side of a cement plant, on the other hand, Biodegradable waste, carry bag, NRPW, plastics, etc., are associated with the supply side of cement plants. Table 2.1 gives a clear picture of the Materials related Terms and Definitions taken from different source documents.

2.3   Process and System-Related Terms and Definitions

Several processes and systems are related to co-processing and associated activities. The process-related terms are associated with the co-processing such as Pre-processing, Alternate use, Calcination, Clinkering, environmentally sound technologies while environmentally sound management, solid waste management (SWM), Transboundary movement, Sustainable Development are system-related terms. Table 2.2 demonstrates the Process and System-related Terms and Definitions.

2.3.1   Legal and Permission-Related Terms and Definitions

The terms, such as Authorization, Consent, extended producer responsibility (EPR), Registration, Prescribed authority, and Tipping (or Gate) Fee are day-to-day relation to the co-processing and associated activities. The permission for the process, compliance to the respective limits and parameters as well as the awareness of each

Table 2.1 Definitions of Materials-related Terms associated with co-processing

S. No	Terms	Definition
1	Alternative Fuels	Wastes with recoverable energy value, used as fuels in a cement kiln, replacing a portion of conventional fossil fuels such as coal. Other terms include secondary, substitute, or waste derived fuels (UNEP, 2011)
2	Alternative fuels and raw materials (AFR)	Inputs to clinker production derived from waste streams that contribute to energy and raw material requirements in the clinker manufacture
3	Alternative raw materials	Waste materials containing useable minerals such as calcium, silica, alumina, and iron, which can be used in the kiln to replace raw materials such as clay, shale, and limestone. Also known as secondary or substitute raw materials (UNEP, 2011)
4	Biodegradable waste	Means any organic material that can be degraded by micro-organisms into simpler stable compounds; (SWM Rules, GOI, 2016)
5	Carry bags	Mean bags made from plastic material or compostable plastic material, used for the purpose of carrying or dispensing commodities which have a self-carrying feature but do not include bags that constitute or form an integral part of the packaging in which goods are sealed prior to use; (PWM Rules, GOI, 2016)
6	Cement	Finely ground inorganic material that, when mixed with water, forms a paste that sets and hardens by means of hydration reactions and processes and that, after hardening, retains its strength and stability under water (UNEP, 2011)
7	Clinker	Intermediate product in cement manufacturing and the main substance in cement. Clinker is the result of calcination of limestone in the kiln and subsequent reactions caused through burning (IFC, 2016)
8	Combustible waste	Means non-biodegradable, non-recyclable, non-reusable, non hazardous solid waste having minimum calorific value exceeding 1500 kcal/kg and excluding chlorinated materials like plastic, wood pulp, etc.; (SWM Rules, GOI, 2016)

(continued)

Table 2.1 (continued)

S. No	Terms	Definition
9	Commodity	Means tangible item that may be bought or sold and includes all marketable goods or wares; (PWM Rules, GOI, 2016)
10	Concrete	A material produced by mixing cement, water, and aggregates. The cement acts as a binder, and the average cement content in concrete is about 15% (GIZ, 2020)
11	Dry waste	Means waste other than biodegradable waste and inert street sweepings and includes recyclable and non-recyclable waste, combustible waste and sanitary napkin and diapers, etc.; (SWM Rules, GOI, 2016)
12	Dust	Total clean gas dust after dedusting equipment. (In the case of cement kiln main stacks, more than 95% of the clean gas dust has PM10 quality, i.e., is particulate matter (PM) smaller than 10 microns) (GIZ & Lafarge Holcim, 2020)
13	Food-stuffs	Mean ready to eat food products, fast food, processed or cooked food in liquid, powder, solid or semi-solid form; (PWM Rules, GOI, 2016)
14	Inert	Means wastes which are not biodegradable, recyclable, or combustible street sweeping or dust and silt removed from the surface drains; (SWM Rules, GOI, 2016)
15	Multi-layered packaging	Means any material used or to be used for packaging and having at least one layer of plastic as the main ingredients in combination with one or more layers of materials such as paper, paper board, polymeric materials, metalized layers or Aluminium foil, either in the form of a laminate or co-extruded structure; (PWM Rules, GOI, 2016)
16	Non-biodegradable waste	Means any waste that cannot be degraded by micro-organisms into simpler stable compounds; (SWM Rules, GOI, 2016)

(continued)

Table 2.1 (continued)

S. No	Terms	Definition
17	Non-Recyclable Plastics Waste (NRPW)	Recyclability of plastics depends on the Polymer groups; 5 out of 7 of the Plastic groups defined by the Resin Identification Codes are never or seldom recycled due to physical degradation under the recycling process, content of various additives like chlorine, etc. The following four criteria need to be in place for a product to be recyclable: 1. The product must be made with a plastic that is collected for recycling, has market value and/or is supported by a legislatively mandated programme 2. The product must be sorted and aggregated into defined streams for recycling processes 3. The product can be processed and reclaimed/recycled with commercial recycling processes 4. The recycled plastic becomes a raw material that is used in the production of new products Innovative materials must demonstrate that they can be collected and sorted in sufficient quantities and must be compatible with existing industrial recycling processes or will have to be available in sufficient quantities to justify operating new recycling processes The definition can be found here: https://waste-management-world.com/a/global-definition-of-plasticsrecyclability-from-international-recycling-associations
18	Other wastes	Means wastes specified in Part B and Part D of Schedule III for import or export and includes all such waste generated indigenously within the country; (HOWM Rules, GOI, 2016)
19	Plastic	Means material which contains as an essential ingredient a high polymer such as polyethylene terephthalate, high density polyethylene, Vinyl, low density polyethylene, polypropylene, polystyrene resins, multi-materials like acrylonitrile butadiene styrene, polyphenylene oxide, polycarbonate, and Polybutylene terephthalate; (PWM Rules, GOI, 2016)

(continued)

Table 2.1 (continued)

S. No	Terms	Definition
20	Plastic sheet	Means Plastic sheet is the sheet made of plastic; (PWM Rules, GOI, 2016)
21	Plastic waste	Means any plastic discarded after use or after their intended use is over; (PWM Rules, GOI, 2016)
22	Refuse Derived Fuel (RDF)	Solid fuel prepared from the energy rich fraction of municipal solid waste after the removal of recyclables (IFC, 2016)
23	Solid waste	Means and includes solid or semi-solid domestic waste, sanitary waste, commercial waste, institutional waste, catering and market waste and other non-residential wastes, street sweepings, silt removed or collected from the surface drains, horticulture waste, agriculture and dairy waste, treated bio-medical waste excluding industrial waste, bio-medical waste and e-waste, battery waste, radio-active waste generated in the area under the local authorities and other entities mentioned in rule 2; (SWM Rules, GOI, 2016)
24	User fee	Means a fee imposed by the local body and any entity mentioned in rule 2 on the waste generator to cover full or part cost of providing solid waste collection, transportation, processing, and disposal services. (SWM Rules, GOI, 2016)
25	Virgin plastic	Means plastic material which has not been subjected to use earlier and has also not been blended with scrap or waste; (PWM Rules, GOI, 2016)
26	Waste	Any substance or object which the holder discards or is required to discard (WBCSD, 2014)
	Waste	The EC Framework Waste Directive 75/442/EEC, Article 1 defines waste as "any substance or object, which (a) the holder discards or intends or is required to discard or (b) has to be treated in order to protect the public health or the environment." Waste material can be solid, liquid, or pasty (GTZ & Holcim, 2006)

Table 2.2 Definitions of process and system-related terms associated with co-processing

Sl. No	Terms	Definitions
1	Alternate use	means use of material for a purpose other than for which it was conceived, which is beneficial because it promotes resource efficiency; (PWM Rules, GOI, 2016)
2	Auto-Ignition Temperature	Auto-ignition temperature is the temperature at which a chemical can burn without an ignition source
3	Bio-methanation	means a process which entails enzymatic decomposition of the organic matter by microbial action to produce methane rich biogas; (SWM Rules, GOI, 2016)
4	Calcination	Heat-induced removal, or loss of chemically-bound volatiles other than water. In cement manufacture, this is the thermal decomposition of calcite (calcium carbonate) and other carbonate minerals that gives a metallic oxide (mainly CaO) plus carbon dioxide (UNEP, 2011)
5	Clinkering	The thermo-chemical formation of clinker minerals, especially to those reactions occurring above about 1,300° C; also the zone in the kiln where this occurs. Also known as sintering or burning (UNEP, 2011)
6	Composting	means a controlled process involving microbial decomposition of organic matter; (SWM Rules, GOI, 2016)
7	Co-processing	The use of suitable waste materials in manufacturing processes for the purpose of energy and/or resource recovery and resultant reduction in the use of conventional fuels and/or raw materials through substitution (UNEP, 2011)
8	De-centralized processing	means establishment of dispersed facilities for maximizing the processing of biodegradable waste and recovery of recyclables closest to the source of generation so as to minimize transportation of waste for processing or disposal; (SWM Rules, GOI, 2016)

(continued)

Table 2.2 (continued)

Sl. No	Terms	Definitions
9	Disintegration	means the physical breakdown of a material into very small fragments; ga. "energy recovery" means energy recovery from waste that is conversion of waste material into usable heat, electricity, or fuel through a variety of processes including combustion, gasification, Pyrolysis, anaerobic digestion & landfill gas recovery"; (PWM Rules, GOI, 2016)
10	Environmentally sound management	means taking all steps required to ensure that the hazardous and other wastes are managed in a manner which shall protect health and the environment against the adverse effects which may result from such waste; (HOWM Rules, GOI, 2016)
11	Environmentally sound technologies	means any technology approved by the Central Government from time to time; (HOWM Rules, GOI, 2016)
12	Flash Point	Flash point is the lowest temperature at which a chemical can vapourize to form an ignitable mixture in air. A lower flash point indicates higher flammability. Measuring a flash point (open-cup or close-cup) requires an ignition source. At the flash point, the vapour may cease to burn when the ignition source is removed
13	Handling	includes all activities relating to sorting, segregation, material recovery, collection, secondary storage, shredding, baling, crushing, loading, unloading, transportation, processing, and disposal of solid wastes; (SWM Rules, GOI, 2016)
14	Heating (calorific) value	The heat per unit mass produced by complete combustion of a given substance. Calorific values are used to express the energy values of fuels, usually expressed in mega joules per kilogram (MJ/kg) (UNEP, 2011)
15	Higher heating (calorific) value (HHV)	Maximum amount of energy that can be obtained from the combustion of a fuel, including the energy released when the steam produced during combustion is condensed, also called the gross heat value (UNEP, 2011)

(continued)

Table 2.2 (continued)

Sl. No	Terms	Definitions
16	Incineration	means an engineered process involving burning or combustion of solid waste to thermally degrade waste materials at high temperatures; (SWM Rules, GOI, 2016)
17	Leachate	means the liquid that seeps through solid waste or other medium and has extracts of dissolved or suspended material from it; (SWM Rules, GOI, 2016)
18	Lower heating (calorific) value (LHV)	The higher heating value less the latent heat of vapourization of the water vapour formed by the combustion of the hydrogen in the fuel. Also called the net heat value (UNEP, 2011)
19	Pre-processing	Alternative fuels and/or raw materials not having uniform characteristics must be prepared from different waste streams before being used as such in a cement plant. The preparation process, or pre-processing, is needed to produce a waste stream that complies with the technical and administrative specifications of cement production and to guarantee that environmental standards are met (UNEP, 2011)
20	Recovery	means any operation or activity wherein specific materials are recovered; (HOWM Rules, GOI, 2016)
21	Recycling	means the process of transforming segregated non-biodegradable solid waste into new material or product or as raw material for producing new products which may or may not be similar to the original products; (SWM Rules, GOI, 2016)
22	Reuse	means use of hazardous or other waste for the purpose of its original use or other use; (HOWM Rules, 2016)

(continued)

personnel involved in coprocessing are prime important aspects in coprocessing business. The Legal and permission-related terms and definitions associated with the co-processing are described in Table 2.3.

Table 2.2 (continued)

Sl. No	Terms	Definitions
23	Sanitary land filling	means the final and safe disposal of residual solid waste and inert wastes on land in a facility designed with protective measures against pollution of ground water, surface water and fugitive air dust, wind-blown litter, bad odour, fire hazard, animal menace, bird menace, pests or rodents, greenhouse gas emissions, persistent organic pollutants slope instability and erosion; (SWM Rules, GOI, 2016)
24	Segregation	means sorting and separate storage of various components of solid waste, namely biodegradable wastes including agriculture and dairy waste, non-biodegradable wastes including recyclable waste, non-recyclable combustible waste, sanitary waste and non-recyclable inert waste, domestic hazardous wastes, and construction and demolition wastes; (SWM Rules, GOI, 2016)
25	Solid Waste Management (SWM)	refers to the supervised handling of waste material from generation at the source through the recovery processes to disposal (IFC, 2016)
26	Sorting	means separating various components and categories of recyclables such as paper, plastic, cardboards, metal, and glass from mixed waste as may be appropriate to facilitate recycling; (SWM Rules, GOI, 2016)
27	Storage	mean storing any hazardous or other waste for a temporary period, at the end of which such waste is processed or disposed of; (HOWM Rules, GOI, 2016)
28	Sustainable development	"Development that satisfies present needs without compromising the ability of future generations to satisfy their own needs," as first defined in the report Our Common Future published by the United Nations Brundtland Commission in 1987 (WBCSD, 2014)

(continued)

Table 2.2 (continued)

Sl. No	Terms	Definitions
29	Transboundary movement	means any movement of hazardous or other wastes from an area under the jurisdiction of one country to or through an area under the jurisdiction of another country or to or through an area not under the jurisdiction of any country, provided that at least two countries are involved in the movement; 33. "transport" means off-site movement of hazardous or other wastes by air, rail, road or water; (HOWM Rules, GOI, 2016)
30	Transportation	means conveyance of solid waste, either treated, partly treated or untreated from a location to another location in an environmentally sound manner through specially designed and covered transport system so as to prevent the foul odour, littering and unsightly conditions; (SWM Rules, GOI, 2016)
31	Treatment	means a method, technique or process, designed to modify the physical, chemical or biological characteristics or composition of any hazardous or other waste so as to reduce its potential to cause harm; (HOWM Rules, GOI, 2016)
32	Utilization	means use of hazardous or other waste as a resource;5 (HOWM Rules, GOI, 2016)
33	Waste management	means the collection, storage, transportation reduction, re-use, recovery, recycling, composting or disposal of plastic waste in an environmentally safe manner; (PWM Rules, GOI, 2016)

2.4 Infrastructure-Related Terms and Definitions

Kiln, Co-incineration plant, Facility, Pre-calciner, Rotary kiln, etc., are the infrastructure which are necessary for effective coprocessing of wastes in cement plant. Table 2.4 described the terms and definitions related to infrastructure for co-processing.

Table 2.3 Definitions of legal and permission-related terms associated with co-processing

Sl. No	Terms	Definitions
1	Authorization	means the permission given by the State Pollution Control Board or Pollution Control Committee, as the case may be, to the operator of a facility or urban local authority, or any other agency responsible for processing and disposal of solid waste; (SWM Rules, GOI, 2016)
2	Basel Convention	means the United Nations Environment Programme Convention on the Control of Transboundary Movement of Hazardous Wastes and their Disposal; (HOWM Rules, GOI, 2016)
3	Central Pollution Control Board (CPCB)	means the Central Pollution Control Board constituted under sub-Sect. (1) of Sect. 3 of the Water (Prevention and Control of Pollution) Act, 1974 (6 of 1974); (HOWM Rules, GOI, 2016)
4	Consent	means the consent to establish and operate from the concerned State Pollution Control Board or Pollution Control Committee granted under the Water (Prevention and Control of Pollution) Act, 1974 (6 of 1974), and the Air (Prevention and Control of Pollution) Act, 1981 (14 of 1981); (PWM Rules, GOI, 2016)
5	Extended Producer Responsibility (EPR)	means responsibility of any producer of packaging products such as plastic, tin, glass, and corrugated boxes for environmentally sound management, till end-of-life of the packaging products; (SWM Rules, GOI, 2016)
6	Local body	for the purpose of these rules means and includes the municipal corporation, nagar nigam, municipal council, nagarpalika, nagar Palikaparishad, municipal board, nagar panchayat and town panchayat, census towns, notified areas and notified industrial townships with whatever name they are called in different States and union territories in India; (SWM Rules, GOI, 2016)
7	Prescribed authority	means the authorities specified in rule 12; (PWM Rules, GOI, 2016)

(continued)

Table 2.3 (continued)

Sl. No	Terms	Definitions
8	Registration	means registration with the State Pollution Control Board or Pollution Control Committee concerned, as the case may be; (PWM Rules, GOI, 2016)
9	State Pollution Control Board	means the State Pollution Control Board constituted under Sect. 4 of the Water (Prevention and Control of Pollution) Act, 1974 (6 of 1974) and includes, in relation to a Union territory, the Pollution Control Committee; (HOWM Rules, GOI, 2016)
10	Tipping (or Gate) Fee	means a fee or support price determined by the local authorities or any state agency authorized by the State government to be paid to the concessionaire or operator of waste processing facility or for disposal of residual solid waste at the landfill; (SWM Rules, 2016)

2.5 Stakeholder-Related Terms and Definitions

The contractor, transporters, waste generators, etc., are very important stakeholders whose work has a great impact on the quality of co-processing and associated activities. The terms and definitions related to stakeholder in the co-processing and associated activities are described in Table 2.5.

Table 2.4 Definitions of infrastructure-related terms associated with co-processing

Sl. No	Terms	Definitions
1	Captive treatment, storage and disposal facility	means a facility developed within the premises of an occupier for treatment, storage and disposal of wastes generated during manufacture, processing, treatment, package, storage, transportation, use, collection, destruction, conversion, offering for sale, transfer or the like of hazardous and other wastes; (HOWM Rules, GOI, 2016)
2	Co-incineration plant	Under Directive 2000/76/EC of the European Parliament and of the Council, any stationary or mobile plant whose main purpose is the generation of energy or the production of material products and which uses wastes as a regular or additional fuel; or in which waste is thermally treated for the purpose of disposal. (GIZ, 2020)
3	Common treatment, storage and disposal facility	means a common facility identified and established individually or jointly or severally by the State Government, occupier, operator of a facility or any association of occupiers that shall be used as common facility by multiple occupiers or actual users for treatment, storage and disposal of the hazardous and other wastes; (HOWM Rules, GOI, 2016)
4	Disposal	means the final and safe disposal of post processed residual solid waste and inert street sweepings and silt from surface drains on land as specified in Schedule I to prevent contamination of ground water, surface water, ambient air and attraction of animals or birds; (SWM Rules, GOI, 2016)
5	Dump sites	means a land utilized by local body for disposal of solid waste without following the principles of sanitary land filling; (SWM Rules, GOI, 2016)
6	Facility	means any establishment wherein the solid waste management processes, namely segregation, recovery and storage, collection, recycling, processing, treatment, or safe disposal are carried out; (SWM Rules, GOI, 2016)

(continued)

Table 2.4 (continued)

Sl. No	Terms	Definitions
7	Kiln	The part of the cement plant that manufactures clinker; comprises the kiln itself, any preheaters and Pre-calciner and the clinker cooler apparatus (UNEP, 2011)
10	Pre-calciner	A kiln line apparatus usually combined with a preheater, in which partial to almost complete calcination of carbonate minerals is achieved ahead of the kiln itself, and which makes use of a separate heat source. A Pre-calciner reduces fuel consumption in the kiln, and allows the kiln to be shorter, as it no longer has to perform the full calcination function (UNEP, 2011)
11	Preheater	An apparatus for heating the raw mix before it reaches the dry kiln itself. In modern dry kilns, the preheater is commonly combined with a Pre-calciner. Preheaters use hot exit gases from the kiln as their heat source (UNEP, 2011)
13	Rotary kiln	A kiln consisting of a gently inclined, rotating steel tube lined with refractory brick. The kiln is fed with raw materials at its upper end and heated by flame from, mainly, the lower end, which is also the exit end for the product (clinker) (UNEP, 2011)

Table 2.5 Definitions of stakeholder-related terms associated with co-processing

Sl. No	Terms	Definitions
1	Brand owner	means a person or company who sells any commodity under a registered brand label. (SWM Rules, GOI, 2016)
2	Contractor	means a person or firm that undertakes a contract to provide materials or labour to perform a service or do a job for service providing authority; (SWM Rules, GOI, 2016)
3	Importer	means a person who imports or intends to import and holds an Importer–Exporter Code number, unless otherwise specifically exempted. (PWM Rules, GOI, 2016)
4	Institutional waste generator	means and includes occupier of the institutional buildings such as building occupied by Central Government Departments, State Government Departments, public or private sector companies, hospitals, schools, colleges, universities or other places of education, organization, academy, hotels, restaurants, malls, and shopping complexes; (PWM Rules, 2016)
5	Manufacturer	means and includes a person or unit or agency engaged in production of plastic raw material to be used as raw material by the producer. (PWM Rules, GOI, 2016)
6	Operator of a facility	means a person or entity, who owns or operates a facility for handling solid waste which includes the local body and any other entity or agency appointed by the local body; (SWM Rules, GOI, 2016)
7	Producer	means persons engaged in manufacture or import of carry bags or multi-layered packaging or plastic sheets or like, and includes industries or individuals using plastic sheets or like or covers made of plastic sheets or multi-layered packaging for packaging or wrapping the commodity; (PWM Rules, GOI, 2016)
8	Street vendor	shall have the same meaning as assigned to it in clause (l) of subsection (1) of Sect. 2 of the Street Vendors (Protection of Livelihood and Regulation of Street Vending) Act, 2014 (7 of 2014); (PWM Rules, GOI, 2016)
9	Transporter	means a person engaged in the off-site transportation of hazardous or other waste by air, rail, road, or water; (HOWM Rules, GOI, 2016)
10	Waste generator	means and includes every person or group of persons, every residential premises and non-residential establishments including Indian Railways, defense establishments, which generate solid waste; (SWM Rules, GOI, 2016)
11	Waste pickers	mean individuals or agencies, groups of individuals voluntarily engaged or authorized for picking of recyclable plastic waste. (PWM Rules, GOI, 2016)

References

GIZ & Lafarge Holcim. (2020). Guidelines on Pre- and Co-processing of Waste in Cement Production. Deutsche Gesellschaft für Internationale Zusammenarbeit (GTZ) GmbH, Holcim Technology Ltd.

GTZ-Holcim. (2006). Guidelines on co-processing Waste Materials in Cement Production. The GTZ-Holcim Public Private Partnership. Holcim Group Support Ltd and Deutsche Gesellschaft für Technische Zusammenarbeit (GTZ) GmbH.

HOWM Rules, GOI. (2016). "Hazardous and Other Wastes (Management and Transboundary Movement) Rules, 2016" notified by MoEFCC, Government of India, on 4th April 2016.

IFC. (2016). Unlocking Value: alternative Fuels For Egypt's cement industry. International Finance Corporation (IFC). www.ifc.org.

PWM Rules, GOI. (2016). Plastic Waste Management Rules 2016 notified by MoEFCC, Government of India on 18th March 2016 and subsequently amended on 27th March 2018.

SWM Rules, GOI. (2016). "Solid Waste Management Rules 2016" notified by MoEFCC, Government of India on 8th April 2016.

UNEP. (2011). Technical guidelines on the environmentally sound co-processing of hazardous wastes in cement kilns. United Nations Environment Programme (UNEP), 11 November 2011. UNEP/CHW.10/6/Add/3/Rev. 1.

WBCSD. (2014). Guidelines for Co-Processing Fuels and Raw Materials in Cement Manufacturing. World Business Council for Sustainable Development (WBCSD), Cement Sustainability Initiative (CSI). www.wbcsd.org.

Part II
Literature Review

Chapter 3
Status Review of Research on Co-processing

3.1 Introduction

Portland cement clinker is a very important compound of modern cements. CO_2 emission during the calcination of calcium carbonate as raw material takes place in cement plants. Reduction of CO_2 emission, the anthropologically caused climate change, is the focus of international initiatives, and hence, finding and development of strong alternatives are the key areas of researchers, policy makers, and plant operators. Operators have been working to use more secondary raw materials and alternative fuels available in close proximity to the plants, whereas the research and developments in cement clinker chemistry show a significant potential for alternatives. For the last two decades, different types of wastes have been seen to work effectively as AFRs with higher thermal substitution rates (TSR) and cost-effective initiatives in cement plants. To achieve these, a significant understanding and improvement in cement clinker chemistry are needed. Baidya and Ghosh analysed the substitution benefit and monetary benefits based on the four constructs in co-processing in cement kiln, namely amount of AFRs co-processed, TSR%, TF, and TR replaced (Baidya & Ghosh, 2019).

Coal is the main and primary fuel used in cement kilns. Presently, the use of alternative fuels in cement kilns is in practice and increasing at a faster rate while a wider range of alternative fuels is being used successfully in gas, liquid, and solid forms (Table 3.1). In 2005, nearly 9 exajoules (EJ) of fuels and electricity were consumed by the global cement industry for cement production (IEA, 2007; Murray & Price, 2008). Energy consumption varies depending on the technology and scale of operations. Nearly 3.2–6.3 GJ energy and about 1.7 tons of raw materials (mainly limestone) per ton are required in the production of Clinker (Rahman et al., 2015). Cement production is an energy-intensive industry that utilizes thermal energy of cost equivalent to 20–25% of the total cement production cost (Madlool et al., 2013) while the estimated lifetime of a kiln used in the cement industry is nearly 30–50 years (Peter & Martin, 2019).

Table 3.1 Different categories of liquid, solid, and gaseous alternative fuels that can be used in the cement industry

Category	Fuels
Liquid Fuels	Used oils, petrochemical waste, paint waste, Tar, chemical wastes, distillation residues, waste solvents, wax suspensions, oil sludge, asphalt slurry, etc.
Solid Fuels	Used tyres, battery cases, paper waste, rubber residues, pulp sludge, dried sewage sludge, plastics wastes and residues, wood waste, domestic refuse, rice husks, refuse-derived fuel (RDF), coconut shell, nut shells, oil-bearing soils, non-cattle feed harvest rejects, etc.
Gaseous Fuels	Refinery waste gas, landfill gas, pyrolysis gas, natural gas, etc.

Long residence time, high temperatures in cement kilns, and several other characteristics in the production of cement make co-processing of wastes as a viable strategic option to the industries and policy makers. The favourable conditions that are created in the clinker production in cement kiln includes high temperature, long residence time, oxidizing atmosphere, ash retention in clinker, alkaline environment, and high thermal inertia. The organic part of the fuel is destroyed under these conditions while the inorganic part as well as the heavy metals are trapped and combined in the product on its own. Since the last three decades, co-processing of wastes has been practised in cement kilns in a few developed countries, namely a number of countries in the European Union, the United States, and Japan. Many developing countries such as India, China, Malaysia, South Korea, and a few other countries in Asia have been practising co-processing to promote the co-processing of wastes in the cement industry. Landfill and incineration are currently the main and most preferred disposal methods of municipal solid waste with a number of many limitations (Yao et al., 2019). The landfill is easy and the cheapest disposal method while the generation of GHG causes harm to the environment and the leaching causes long-term secondary environmental pollution affecting soil fertility, contamination of aquifer and waterbodies, and the flora–fauna with no recovery of resources.

The use of alternative fossil fuels in several cement plants in European countries saves fossil fuels to a greater extent. As reported in Cembureau, 1999, from 1990 to 1998, the savings in fossil fuels in European cement plants may be equivalent to 2.5 million tons of coal per year. During 1990 and 1998, the AFR use in cement plants in different countries were France @ 52.4%, Switzerland @ 25%, Great Britain @ 20%, Italy @ 4.1%, Belgium @ 18%, Germany 15%, Czech Republic @ 9.7%, Sweden @ 2%, Poland @ 1.4%, Portugal @1.3%, and Spain @1% (Mokrzycki et al., 2003).

3.2 Feedstock and Raw Materials for Co-processing

Co-processing involves wastes as alternative fuels in cement kilns which are demonstrated in Table 3.1 and Fig. 3.1. It can be observed from Table 3.1 and Fig. 3.1 that there exists a wider scope of using various waste materials in liquid, solid, and gaseous

Fig. 3.1 Classified wastes used as an alternative fuel

states in the cement kiln as alternative materials. Hence, if co-processing is effectively established in the country, a huge amount of waste can be recycled protecting natural resource depletion and reducing GHG emissions with cost-effective cement production.

The municipal wastes consist of several fractions which are combustible but non-recyclable, for example, soiled paper, soiled cloth, soiled and contaminated plastics, pieces of leather, rubber, tyre, multi-layer plastics and other materials, different types of packaging materials, polystyrene (thermocol), and wood, and many others are disposed of to the landfill.

Most of these materials are destined to landfill sites which created lots of pollution as well as a huge amount of resources are depleted. Refuse-derived fuel (RDF) can be produced using these portions of wastes. RDF is used as an alternative fuel in various industries and as AFR in the cement plants helping to promote the circular economy concept. The raw materials and feedstock for the co-processing can be made available from different sources. Some of the possible sources have been shown in Fig. 3.2 which includes different types of industries, waste treatment plants including incineration units, farming and breeding units, construction and demolition activities, wastewater purifiers, food processing units, and many more including the sources of municipal waste streams.

The effectiveness of utilization of any specific feedstock depends on different characteristics of the materials. The key factors are Ash contents, moisture contents, gross calorific value, chlorine, and Sulphur contents. Table 3.2 demonstrates the value of key parameters of different feedstock materials used in different cement plants using co-processing of wastes as AFR. It is observed from Table 3.2 that most of the feedstock used for alternative fuels have higher value of gross calorific value (GCV) than RDF at a level of 2000 to 3000 kcal/kg while Carbon black/powder, Footwear, Low-moisture Plastic waste, Tyre chips, and Resin waste have two to three times higher GCV. The availability of these materials depends on seasonal variation whereas the availability of refuse-derived fuel (RDF) is more or less consistent.

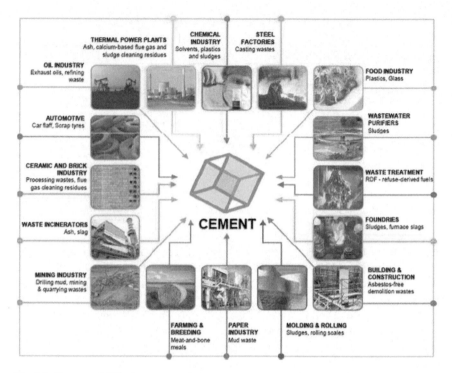

Fig. 3.2 Sources of different raw materials for co-processing in cement kiln

Hence, the cement plants have to depend on the RDF most of which is not very cost-effective than using another feedstock.

This may be noted that a few of the materials mentioned here are available at free of cost or at a rock bottom price which brings the cost of fuel and raw materials to lower levels in cement plants. In Fig. 3.3, a typical modern cement plant, Carthage Cement, in Tunisia and feedstock transportation and preparation in RDF facility are shown. Hazardous waste has diverse characteristics while strict environmental regulations and scarcely available resources such as land, energy, and finance have made its management more complex. Varied handling protocols, alternatives for treatment, and disposal are available for specific types of wastes (Nema & Gupta, 1999). There are several treatment methods for hazardous waste, e.g., acid–base neutralization, incineration, chemical fixation/ solidification, co-processing in cement plants, etc., used prior to final disposal (Misra & Pandey, 2005; Millano, 1996; Baidya et al., 2017a).

The estimated MSW generation in India is to be 140 million tons by 2025 from 62 million tons in 2016. Effective management of MSW can prevent and control human health issues, environmental pollution, and resource depletion. A major portion of MSW, 55–80%, is disposed of to landfill and dump yards in most of the developing countries. Nearly 32% of total primary energy is derived from biomass in rural areas

Table 3.2 Value of key parameters of alternate fuels used in cement plants

Sl	Alternative fuel	Ash content (%)	Moisture (%)	Gross Calorific Value (GCV) (Kcal/Kg)	Cl (%)	S (%)
1	Carbon black/powder	1	15	6000–6300	0.4	2
2	Footwear	5–10	2–5	5400–5700		
3	Low-moisture Plastic waste	5–10	5	5000–6000	–	0.2–0.4
4	Tyre chips	5–10	5–10	4500–5500	0.7	1.5
5	Resin waste	10	15	4000–5000	0.5	–
6	Cashew nut shell	8–15	5–10	3500–4000		
7	Plastic waste	5–10	35–40	3200–3500		
8	Paint sludge	15–20	20–30	3000–3500		
9	High-moisture plastic waste	50	5	3000	–	–
10	Rice husk	10–15	2–5	3000–3500		
11	Refuse-derived fuel (RDF)	15–30	10–30	2000–3000	0.4–0.7	0.2–0.3

Table 3.3 Spent pot liner waste: estimated projections in India. (*Source* CII, 2016)

S.No	Parameter	Existing 2015	Anticipated 2025	Unit
1	Calorific value	4000	4000	Kcal/kg
2	Quantitty of SPL generated[14]	0.04	0.07	Million TPA
3	Quantity already in secured landfills	0.09	0.09	Million TPA
4	Total quantity	–	0.16	Million TPA
5	Quantity of SPL available for Co-processing	–	0.14	@70%
6	Energy Generated from SPL	–	0.057	Million TOE
7	% Energy from Spent pot liner on AF	–	0.81	%
8	% Energy from spent pot liner on Total thermal energy for cement plant	–	0.20	%

in India and in major developing countries. As per the Ministry of New and Renewable Energy (MNRE) in India, 120 *million tons*/annum of biomass is available as surplus which is a potential feedstock for co-processing in cement plants. Used tyres having higher GCV are a potential alternate fuel generated at the rate of 0.83 million tons per year in India which can be better managed in cement kiln for co-processing. About 0.6 million tons of incinerable hazardous waste are generated annually from 41,523 hazardous waste generating units in India as of 2019 as per the Central Pollution Control Board, India. New Hazardous Waste Management Rules 2016 consider

Fig. 3.3 A modern cement plant: Carthage Cement, Tunisia. (*Source* Amine Abdelkhalek, entrant to the Global Cement Photography Competition and feedstock transportation and preparation for RDF facility)

co-processing as a preferred option for the treatment and disposal of hazardous waste. Figure 3.4 shows photographs of MSW landfill, biomass, tyre waste, and hazardous wastes which are potential feedstock for co-processing in cement plants. The energy in percentage of Total Substitution Rate (TSR) available from different fuels is demonstrated in Fig. 3.5 and the percent energy using different fuels for TSR is shown in Fig. 3.6.

Fig. 3.4 **a** Photographs of a Waste dump site; **b** Agricultural Wastes; **c** Tyre Wastes, **d** Hazardous Wastes. Potential feedstock for RDF/co-processing facility

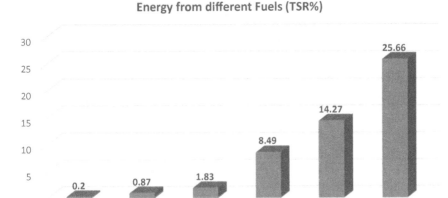

Fig. 3.5 Energy in percentage of TSR available from different fuels. (*Source* CII, 2016)

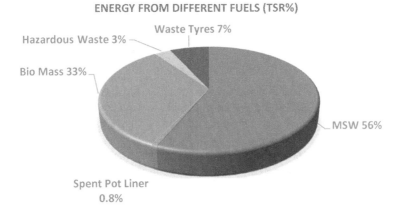

Fig. 3.6 Percent energy using different fuels for TSR. (*Source* CII, 2016)

3.2.1 Municipal Solid Waste (MSW)

Contents of municipal solid waste (MSW) are defined in the national rules of the respective country while in India the definition is governed by SWM Rules 2016. Currently, the estimated 62 million tons of MSW are generated annually. "Swachhata Sandesh Newsletter" by the MoHUA reported, in January 2020, that 147,613 tons/day of solid waste are generated, from 84,475 wards in India. The report by the then Planning Commission in (2014) estimates the waste generation as 276,342 tons per day (TPD) of waste by 2021, 450,132 TPD by 2031, and 1,195,000 TPD by 2050. The quantity of plastic waste generation is also increasing which is another significant challenge. India generates 9.4 million tons per annum of plastic waste. The MoEFCC estimates that only 75–80 percent of the total MSW gets collected and 22–28 percent is treated (Press Information Bureau, 2016).

3.2.2   Spent Pot Liner Waste

In primary aluminium production, carbon anodes and electrolyte bath and carbon cathodes are used. Alumina ($Al2O3$) dissolved in cryolite ($Na3AlF6$) is contained in the electrolyte at 960 °C temperatures (Ismael Vemdrame Flores et al., 2019; Gunasegaram & Molennar, 2015). At an interval of 3–8 years of operation, the generated carbon cathode "potlining" is removed from the shell and the shell is re-lined. The resulting "potlining" material as removed is called SPL (spent pot lining). SPL is a hazardous waste to be disposed of in secured storage landfilling (SLF) sites. SPL waste contains 4000–5000 kcal/kg of gross calorific value (GCV). SPL is a preferred alternative fuel in cement kilns that can be easily used in cement plants through co-processing. SPL is available in India and in many other countries.

3.2.3   Hazardous Waste

Usually, hazardous waste is burnt in incinerators in the common hazardous waste treatment, storage, and disposal facilities (TSDFs) or at captive treatment facilities at waste generating unit, and a portion is disposed of in sanitary landfills as per the rules (Ghosh, 2017). The energy contents in the hazardous waste can be used as AFR in a cement kiln. All the countries generate a huge amount of hazardous waste though the disposal of hazardous waste is carried out strictly as per national hazardous waste rules [In India, Hazardous and Other Wastes (Management and Transboundary Movement) Amendment, Rules, 2016 latest amended in 23/4/2019] because of its adverse impact on environment and health. In India, as per the Central Pollution Control Board (CPCB) report 2019, nearly 41,523 units generate 7.90 Million Tons of hazardous waste per annum out of which the incinerable and recyclable portion is nearly 4.58 Million Tons per annum. Table 3.4 gives the estimated projection of hazardous wastes in India.

3.2.4   Biomass Waste

Biomass contains sawdust, rice and paddy husk, wastes from food industry waste, and many others. In the rural areas in Asia, Africa, and Latin America, biomass is used as the primary energy source derived from biomass. Biomass produces energy via combustion and after converting it to various forms of biofuel. Biomass is utilized

Table 3.4 Hazardous waste: estimated projections. (*Source* CII, 2016)

S.No	Parameter	Existing 2011	Anticipated 2025	Unit
1	Incinerable Hazardous waste	0.6	1.02[*]	Million TPA
2	Average calorific value[16]	3000	3000	Kcal/kg
3	Percentage of quantity available for co-processing		80	%
4	Quantity available for co-processing		0.81	Million TPA
5	Energy from Hazardous waste		0.24	Million TOE
6	% Energy from HW on AF		3.46	%
7	% Energy from HW on Total thermal energy for cement plant		0.87	%

Table 3.5 Biomass waste: estimated projections. (*Source* CII, 2016)

S.No	Parameter	Existing 2015	Anticipated 2025	Unit
1	Biomass generation in India19	500	–	Million TPA
2	Surplus biomass availability	120	120	Million TPA
3	Average Calorific value		3000	Kcal/kg of material
4	Available for co-processing		6.5	%
5	Quantity from Biomass for co-processing		8	Million TPA
6	Energy from Biomass		2.40	Million TOE
7	% Energy from Biomass on AF	–	33.97	%
8	% Energy from Biomass on total thermal energy for cement plant	–	8.49	%

in the production of bioenergy and thus helps in the bioeconomy. In India, the agricultural and forestry residues contribute to about 120–150 million tons per annum of surplus biomass availability as estimated by the Ministry of Non-Renewable Energy (MoNRE) and more than 6.5% is used for co-processing for the production of cement. There is a huge scope of utilization of biomass in co-processing in cement plants in India and many other countries. Table 3.5 shows the projection of biomass in India.

3.2.5 Tyre Waste

Globally, as estimated, nearly 13.5 million tons of tyres are scrapped each year; 40% of which come from emerging markets such as China, India, South America, Southeast Asia, South Africa, and Eastern Europe. In the US alone, exports of waste

Table 3.6 Tyre waste: estimated projections. (*Source* CII, 2016)

S.No	Parameter		Anticipated	Unit
1	Used tyres availability	0.83	1.32^{23}	Million TPA
2	Average Calorific value		6500	Kcal/kg
3	Percentage of quantity available for co-processing		60	%
4	Quantity available for co-processing		0.79	Million TPA
5	Energy from Tyre waste		0.51	Million TOE
6	% Energy from Tyre waste on AF	–	7.33	
7	% Energy from Tyre waste on total energy for cement plant	–	1.83	

tyres amounted to almost 140,000 tons/year from 2002–2011. Globally, over 1.6 billion new tyres are produced and nearly 1 billion waste tyres are generated and only 100 million tyres are processed and recycled every year. India discards about 275,000 tyres a day whereas these are pyrolyzed in 637 tyre pyrolysis units in 19 states of the country as of July 2019. In the European Union, the End-of-Life Vehicle Directive (2000/53/EC) governs and stipulates the separate collection of tyres from vehicles dismantlers and encourages the recycling of tyres and other materials. India Tyre Manufacturing Association (ATMA) reported that 0.83 million tons of used tyres were generated annually in 2011–2012. Tyre waste is a superior alternate fuel having a calorific value of > 6500 kcal/kg to use in cement co-processing. Tyre-derived fuel (TDF) was first used in cement kilns, in Germany, in the 1970s; subsequently, 50% of the used tyre in the USA are converted into TDF for the cement industry. Table 3.6 shows the estimates of tyre wastes in India for 2025.

3.2.6   Summary of Energy Available From Different Waste Streams

After assessing the existing quantity and future forecasting of different waste generation and waste-derived fuels, it may be concluded that 25% alternate fuel substitution is possible in the Indian cement industry by 2025. TSR @ 25% will benefit individual unit and support the national initiative of achieving intended nationally determined contribution (INDC). Table 3.7 shows the percentage share of alternate fuel and total thermal energy in India.

Table 3.7 Percent share of alternate fuel and total thermal energy in India. (*Source* CII, 2016)

S.No	Waste streams	% share on AF	% share on total thermal energy
1	MSW	57.07	14.27
2	Spent Pot Lining	0.81	0.20
3	Biomass	33.97	8.49
4	Hazardous waste	3.46	0.87
5	Tyre waste	7.33	1.83
	Total		**25.66**

3.3 Guiding Principles for Co-processing of Waste in Cement Plant

The guidelines prepared by GTZ-Holcim Public Private Partnership (GIZ-Holcim, 2006) proposes twenty-two principles focussing on different aspects. It contains legal aspects with seven principles, operational aspects with five principles, occupational health and safety (OH&S) aspects with five principles, and communication and social responsibility aspects with five principles presented in Table 3.8.

3.4 Previous Research on Co-processing and Allied Fields

Researchers have been studying different aspects of recycling and utilization of MSW incineration ash and concluded that both fly ash and bottom ash from MSWI can be used as raw materials for Portland cement. Research results have also been found production and testing of Eco-cement and Eco Concrete using different types of wastes, namely MSW, fly ash, bottom ash, etc. (J.R. et al., 2008; Saikia et al., 2007; Hanehara, 2001; Kikuchi, 2001; Lam et al., 2010, 2011; Kyle A. Clavier, 2019). The popularity of replacing traditional raw materials with wastes and residues hasbeen increasing, and thus, research on this subject is also gaining popularity. In the US context, researchers have conducted several studies (Guo et al., 2016; Pan et al., 2008; Saikia et al., 2007; Hanehara, 2001), but the information on the use of MSW incineration ash into cement production is still inadequate. Fly ash has a perceived risk of heavy metal and chloride leaching while bottom ash does not have those possibilities. Most studies on MSW incineration ash-amended cement focuses on the use of MSWI fly ash, while the bottom ash (BA) is the preferred one for recycling into US cement production.

There are hundreds of literature which are available. Table 3.9 gives glimpses of the study presented by selected researchers in the literature in the span of the last twenty-five years.

Table 3.8 The excerpts of the guiding principles for co-processing waste materials in cement production

Principle number	Statement of principles	Expansion of statement of principles
Legal Aspects:		
1	"An appropriate legislative and regulatory framework shall be set up"	Co-processing including AFR implementation and marketing should be a part in the national legislative instrument concerning environmental protection and waste management
2	"Baselines for traditional fuels and raw materials shall be defined"	Baseline data of initial review of environmental impacts of AFR derived from environmental impact assessments (EIA) is to be used to compare the same with virgin fuel and primary raw materials. Adequate and robust monitoring and control of inputs, outputs, and emissions during the co-processing should be established
3	"All relevant authorities should be involved during the permitting process"	Establish a system of effective open, consistent, and continuous internal and external communication to interested parties/stakeholders to build credibility with the authorities to evaluate the option of co-processing. Setting up of a community advisory panels including the authority/ies is proposed to facilitate multipurpose communication channel, know-how exchange, and advocacy of application of best available technology (BAT)
Environmental Aspects of Cement Production and AFR pre-processing		
4	"Rules must be observed"	The adverse impact of emissions from the stack of cement kiln should be minimized while using AFR by (1) feeding AF into the high-temperature zones (i.e., via the main burner, mid kiln, transition chamber, secondary (riser duct) firing, and pre-calciner firing); (2) feeding alternative raw materials with elevated amounts of volatile matter (organics, sulphur); and (3) developing cement production lines having feeding operation effective. It must filter the dust directed to the cement mills

(continued)

Table 3.8 (continued)

Principle number	Statement of principles	Expansion of statement of principles
5	"Emission monitoring is obligatory"	Emissions monitoring needs to be complied with the national regulations and agreements and company corporate rules, if any. Quality control of input materials mainly the feedstock including AFR for co-processing must be carried out with a higher confidence level
6	"Pre-processing of waste is required for certain waste streams"	Uniform flows of raw material and fuel with respect to quality and quantity in the kiln are required for optimum operation. Pre-processing is carried out for certain types of waste for co-processing
7	"Environmental impact assessments (EIA) confirm compliance with environmental standards"	Environmental and business risk assessments as well as material flux and energy flow analyses need to be carried out for identifying system weaknesses and optimization of resources use, respectively
Operational Issues		
8	"The sourcing of waste and AFR is essential"	Traceability from the receiving to final treatment of AFR at pre-processing and/or co-processing facility has to be maintained to minimize undesired emissions and operational risks and maximize final product quality
		Uniform waste stream with quality and delivery criteria should be ensured through effective contractual business agreements with appropriate stakeholders. SOP (Standard Operating Procedure) must be developed for the introduction of the new waste for source qualification test procedure and acceptance criteria to be followed. Non-conforming waste categories for co-processing must be quarantined and not be used in the process

(continued)

Table 3.8 (continued)

Principle number	Statement of principles	Expansion of statement of principles
9	"Materials transport, handling, and storage must be monitored"	Effective SOP and appropriate machineries/equipment at different stages, namely movement, handling, solid and liquid wastes storage, and AFR should be established and maintained in the plant and should comply with applicable regulatory requirements. Effective design to minimize fugitive dust emissions is required for conveying, dosing, and feeding systems. Effective emergency response plan including response to spill and emission of toxic/ harmful vapours considering start-up and shutdown also should be developed in place with regular mock drill involving all employees
10	"Operational aspects must be considered"	Introduction points of the AFR should be determined considering the characteristics of the AFR and accordingly fed to the kiln system. Plant health monitoring should be effective. The AFR feeding strategies adopted have to be recorded and must be accessible to operators for all conditions of operations, e.g., start-up, shutdown, or emergency/upset conditions of the kiln
11	"Quality control system is a must"	Quality Assurance of process and products must be governed by QAP (Quality Assurance Plans) in pre-processing or co-processing of wastes and AFR supported by resource provision like application documents including SOP, machineries/equipment, and trained manpower. Appropriate protocols for non-conforming AFR/Wastes at incoming, in-process, and outgoing stages must be followed and communicated to operators

(continued)

Table 3.8 (continued)

Principle number	Statement of principles	Expansion of statement of principles
12	"Monitoring and auditing allow transparent tracing"	Internal Auditing procedure should be developed. Periodic regular internal/external auditing by trained personnel/consultants have to be conducted at a specified interval in pre- and co-processing facilities. Audit reports to be communicated and non-compliances are to be closed with timeline
Occupational health and safety (OH&S) issues		
13	"Site suitability avoids risks"	Different types of risks should be avoidable at the site. The following may be considered while assessing and mitigating OHS Risks: (a) Proximity to human settlement, forest, and impact of logistics/transport; (b) good infrastructure to prevent or protect from impacts from emission, dust, water pollution, fire, etc., and (c) trained manpower to handle and process AFR with minimized risks
14	"Safety and security"	Effective organization structure with a responsibility of risk manager for handling safety and security arrangement and execution must be developed, established, and available at site
15	"Documentation and information are must"	Documented OHS management system including the compliance with the system must be established, communicated, and available to employees and authorities before starting co-processing activity
16	"Training should be provided at all levels"	A system should exist to train and retrain and enhance competence of the existing and new employees, contractors, and Management personnel on co-processing. Field visits to existing facilities should be a part of orientation training for all employees to be engaged in the co-processing and allied activities

(continued)

Table 3.8 (continued)

Principle number	Statement of principles	Expansion of statement of principles
		Outsiders have to undergo an induction session. Trainings on risks and mitigation plans are mandatory
17	"Emergency and spill response plans"	Refer to Principle 9 for Effective emergency and spill response as well as DMP (Disaster Management Plan) taking industries and the authorities in neighbourhood should be established and effective
Issues related to Communication and Social Responsibility		
18	"Openness and transparency"	External communication channel with transparency in sharing all necessary information with stakeholders should exist to understand the purpose of co-processing, the context, the function of stakeholders, and decision-making procedures. Sharing of experiences/practices is encouraged
19	"Credibility and consistency"	Build credibility by being open, honest, and consistent
20	"Cultivating a spirit of open dialogue, based on mutual respect and trust"	Refer to Principle 3 on internal and external communication system. A system of seeking feedback and consultative dialogue with internal and external stakeholders should be established
21	"Cultural sensitivity"	Local cultural environments and concerns of sensitivity have to be considered while planning activities and operations with target-orientation and truthful
22	"Continuity"	Start early, and once you start, never stop

Modified from the Guidelines on co-processing Waste Materials in Cement Production, 2006

Table 3.9 Selected literature review on co-processing of waste and allied areas of research

Year	Paper title and Remarks
Rigo and Rigo Associates (1995)	Combustion efficiency (CO and total hydrocarbons) is usually measured to analyse the characteristics in cement kilns and this study did not find any strong relationship between the combustion method and dioxin emission
M.A. Trezza and A.N. Scian (2000)	In the study, ashes in small amount from pyrolysis of used oil from cars were added in the clinkering process of Portland cement. The study simulates the burning process in an industrial furnace using up to 30% of this waste fuel to check the feasibility of using the industrial wastes as alternative fuel in the cement manufacturing
Richard Bolwerk (2004)	The study involved burning of 50–75% alternative combustibles wastes (18–25 MJ/kg) and found that the pollutants are burnt safely when liquids are screened and the solid waste-derived fuels are spread in the gas flow. The study also dealt with the emissions of chlorinated compounds like PCB and dioxin, in cement kilns
D. Ziegler et al. (2006)	A set of guiding principles applicable for co-processing is presented with certain recommendations and a few country-specific experiences as guidelines for stakeholders and policy makers for waste management and cement production
Michae leliasboesch et al. (2009)	The study presented a technology-specific modular LCA model for cradle-to-gate assessments of clinker production which has been validated in two case studies. It shows the environmental impact from waste co-processing and respective production technology
Da-Hai Yan et al. (2010)	The study brings forward some variability found in the Chinese cement sector, considering the small inefficient and polluting plants to modern dry pre-calciner kilns with updated performance. The study reported a general lack of co-processing knowledge, poor capabilities, and infrastructure in the Chinese cement industry irrespective of scales of operations

(continued)

Table 3.9 (continued)

Year	Paper title and Remarks
Yufei Yang et al. (2011)	The study was carried out on concrete using the cement produced from co-processing of hazardous wastes as AFR in China and assessed the release of heavy metals from concrete in leaching tests. The results show that the heavy metals contents are much greater than leachability, while the ratio of total amount to leachability was dependent on the type of heavy metal
Moses P.M. Chinyama Maximino Manzanera (Ed.) (2011)	It reviews the associated environmental and socio-economic benefits of using alternative fuels (AF) in cement production. It reviews the challenges of conversion to AF including combustion characteristics, their effect on cement production and quality, and methodologies of avoiding negative effects on cement as the final product
Wendell de Queiroz Lamas et al. (2012)	The article deals with co-processing introducing fuel waste from different industrial activities. It also focuses on the reduction of environmental liabilities if discarded in inappropriate places
Lei Wang et al. (2012)	The study deals with the incorporation of cadmium into clinker and the stabilization rate of cadmium during the clinkerization process
LI Chun-ping (2012)	The study presented some results of experiments on co-processing. After the addition of alternative fuels in cement kiln, the rates of emission and the concentrations of TSPs, HCl, HF, SO2, and CO in the flue gas were enhanced and volatized while there was no significant change observed for NOx. Similarly, Potassium and Chlorine contents were increased to some extent but Sulphur in hot raw materials remained unchanged. There are advantages observed using alternative fuels. For instance, Compressive strength, flexural strength, water demand for normal consistency, and the surface area of clinker were reduced. Based on the clinker saturation, initial and final setting times were increased

(continued)

Table 3.9 (continued)

Year	Paper title and Remarks
Bahareh Reza et al. (2013)	This study investigates impacts on environmental and economic benefits of RDF production. The research concluded that the Cement manufacturing is a potential destination for RDF where conventional fossil fuels may be replaced with less energy-intensive fuel, like RD
Dipl.-Ing. Sebastian Spaun et al. (2015)	In 2013, more than 600,000 t/a of alternative raw materials were used in the Austrian cement industry. The use of alternative raw materials conserves natural resources while the residues from other industrial processes can be utilized in an environmentally friendly manner. The Association of the Austrian Cement Industry (VÖZ) commissioned the renowned German Research Institute of the Cement Industry (FIZ) to develop recommendations for action for the use of alternative raw materials in the cement production
Rahul Baidya et al. (2016a)	The study concluded that co-processing is a sustainable energy and material recovery process. It addressed the issues and challenges associated with sustainable management of industrial wastes based on experience from case studies on Indian cement industries using co-processing
Parlikar U et al. (2016)	The study based on the results of 22 co-processing trials evaluated the effects on different parameters such as emissions, process, and product quality
Baidya R et al. (2016b)	The study analyses the sustainability of co-processing of industrial waste in cement kiln using three case studies. The research presented the economic and environmental gain achievable using industrial waste as AFR in Indian cement kilns and also identified the challenges and issues for effective implementation

(continued)

Table 3.9 (continued)

Year	Paper title and Remarks
John R. Fyffe et al. (2016)	The recycling rates for MSW are currently over 30% in the U.S. The residue, 5 to 15% of total recycled material, contains non-recycled plastics and fibre of high-energy content and is disposed of to landfills. This study explored and reported different benefits with respect to saving in energy, environmental protection, and trade-offs of conversion of NRPW and fibres into 118 Mg of SRF successfully combusted in cement kiln and significantly reduced CO_2 emissions
Baidya R. et al. (2017b)	The study identifies issues and challenges in the supply chain framework of co-processing route in India and its influencing factors
Mohamed M. Elfaham, and Usama Eldemerdash (2018)	The study carried out the analysis of waste material co-processing using modern chemical analytical methods in an Egyptian cement plant to validate for co-processing while it shows favourable results and tracesof some heavy metals and other hazardous elements with a concentration within acceptable limits
Jolanta Sobik-Szołtysek et al. (2019)	The study observed a substantial reduction in the carbon dioxide emissions by substituting conventional fuels with alternative fuels. The experiment was conducted with sewage sludge as an alternate fuel. The adequacy of the chemical composition of sewage sludge was tested to assess its potential as raw material for clinker production, focussing to reduce the demand for natural fossil fuels
Kyle A. Claviera et al. (2019)	The study investigates the associated risk and performance of cement using MSW incinerator bottom ash in the cement kiln as AFR. It observed an industrial-scale beneficial of the project. When 2.8% kiln feed replacement made the incinerator bottom ash as AFR, the physical and environmental performance remains nearly unchanged
Muhammad Shoaib Ashraf et al. (2019)	The study explored to produce ecological cement (eco-cement) at a synthesis temperature of 1100° using incineration energy and residues. The eco-cement achieved structural binding strength on reaction with CO_2. The eco-cement produced using residues (100%) exhibits low early strength by carbonation activation and was more latent hydraulic

(continued)

Table 3.9 (continued)

Year	Paper title and Remarks
Baidya, R., and Ghosh, S. K. (2019)	The amount of AFRs co-processed, TSR%, TF, and TR replaced and substitution benefit in terms of monetary value were identified while analysing the potential reduction of carbon footprint in co-processing
Baidya, R. et al. (2019)	The research carried out a cost–benefit analysis in a pilot study with co-processing of Blast furnace flue dust in cement kiln
Angela J. Nagleb et al. (2020)	The study explored the possibility of using Irish wind turbine blades as AFR in Co-processing in cement plant though co-processing is not being carried out now in Ireland presently. Co-processing in Ireland may be the least impactful, because of material substitution and reduction in transboundary material transportation between Ireland and Germany
Yeqing Li et al. (2020)	The researchers tested for dioxins and detected the trace in solid and gas samples from main points of preheater. The study concluded that the dioxins absorption is at a rate of 91.6% and capture from gas depends on raw mill and bag filter
Michael Hinkel et al. (2019)	The document gives updated guidelines on pre- and co-processing of waste in the cement industry. It presented know-how and practical experiences gained in implementing pre- and co-processing since the first edition (2006) of guidelines that served as a reference document in international agreements (e.g., Basel Convention for Hazardous Waste Treatment) and adaptation of various national guidelines
Baidya, R. et al. (2020)	The article studied co-processing of industry trade rejects in cement plant using experimental trial and environmental and operational sustainability of the process and looked into economic potential of AFR utilization substituting traditional fuel and raw materials
Gisele De Lorena Diniz Chavez et al. (2021)	The study observed that waste streams for RDF production are under varied uncertainties on implementation of related policies in Espírito Santo, Brazil. It also studied the cost–benefit–risk and trade-off assessment and concluded that substitution of fossil fuels by RDF is beneficial to both cement industry and environment

3.5 Co-processing Supply Chain Issues and Challenges

Using AFR in co-processing in the cement kiln has lots of benefits as well as challenges. Table 3.10 demonstrated Issues and Challenges in the supply chain.

Table 3.10 Issues and Challenges through the supply chain of AFR

Supply Side	In process	Outbound demand side
Consistent supply of AFR	Problem in burning and operational stability, reduction in thermal efficiency (If feed size is >50 mm)	Market demand
Sustainable Supply Chain of AFR	Handling as well as additional heat is required for drying, reducing thermal efficiency (If Moisture content is >15%)	Outbound Logistics
Inbound Logistics	Not suitable for direct feeding, required additional cost for conditioning, and causes increase in specific power (Low Calorific value <2500 kcal/Kg)	Regulated cement price
Quality consistency of AFR	Effect of minor component on clinker quality (Proper selection of AFR/RDF)	Quality of clinker/cement
Adherence to contractual agreement by the supplier	Adherence to pre-processing parameters	Cost of production and customer price
AFR quality and material variation	Improper combustion and high material size	
Right equipment for environmental monitoring	Compliance to emission standards	
Seasonal variation in supply of AFR with high GCV	Preheater Oxidized atmosphere. Improper combustion results in clinker output restriction	
Specialized Infrastructure in AFR lab	Analysis of large variety of materials	
Supply of quality Refractory	High Chloride/alkali damages the refractory	

3.6 Sustainable Development and Sustainability Index of Recovery (SIR) of Different Waste Treatment Technologies

The waste by separate streams is a need of effective recycling hence also the requirements of the recycling industries which helps in the internal value market which exists for those wastes. If waste streams with potential market value are segregated, the residual fraction (or stream) is separated to manage the processing of valuable wastes more efficiently with maximum resource efficiency. Most of the waste treatment processes need segregation of wastes and deal with specific types of wastes. It has been observed by researchers that co-processing of wastes in cement plants may be considered as the most sustainable process than any other if considered the four pillars of sustainability. The study revealed (Baidya et al., 2017a) a number of aspects that provide ample indication of the sustainability of co-processing in India. The pilot study carried out in the study indicates a number of issues, which if addressed can effectively make co-processing the most viable waste disposal and management technique in India. The utilization factor of waste in co-processing is maximum because the waste feedstock is spent in the recovery of energy and the rest for the conversion to clinker make the utilization 100%. In case of any other energy recovery processes, such as incineration and waste-to-energy process, the residue as bottom ash and fly ash produced range from 15 to 25%. The recovery is more as the residence time of wastes burning in the furnace is nearly 14 s seconds in co-processing whereas for other processes it varies from 60 s and more. The co-processing can take up a wide variety of different types of waste. Considering the impact on environment, the emission levels are controlled very efficiently as it is a sub-process in the cement manufacturing. Cement plants have their own effective emission monitoring system. There is not much additional cost in the cement plant for utilization of wastes while the WtE plant of incineration plants needs fresh investment which is much more than the case of co-processing in cement plant. The whole supply chain of waste in the co-processing including the pre-processing units needs skilled personnel which generated the scopes of employment. Hence, co-processing has lots of advantages over other processes in consideration of operational, environmental, economic, and social aspects. The assessment of sustainability index of recovery (SIR) of different waste management treatment technologies was carried out and found that co-processing is the most sustainable one. Table 3.11 shows the overall assessment of the sustainability of the technologies, considering all the possible factors in the four pillars of sustainability (Baidya, 2019). The study revealed that co-processing is the most sustainable technology for energy and resource recovery with a sustainability index of recovery (SIR) of 26. The study considered pyrolysis, incineration, gasification, and co-processing of wastes. While assessing the sustainability of these processes, several factors have been considered under four pillars of sustainability, namely operational, economic, environmental, and social requirements, from literature, and convert them into sustainability ranks, 1, 2, 3, and 4 and thus assessing the overall sustainability.

Table 3.11 Sustainability index of recovery (SIR) of waste treatment technologies

Factor Serial	Factors for Sustainability Index of Recovery	Waste Recycling by Pyrolysis	Waste Recycling by Incineration	Waste Recycling by Gasification	Waste Recycling by Co-processing
Operational Requirements for sustainability					
O1	Pre-Treatment/ Preparation of feed stock before final processing	Required (H) (Bosmans et al., 2013)	Required (M to H) (Bosmans et al., 2013)	Required (H) (Arena, 2012)	Required (H) (Chatziaras et al., 2014)
	Sustainability Rank	4	1	4	3
O2	Specific Emissions control	Required (M); Less treatment to meet emission limits (Samolada and Zabaniotou, 2014)	Required (H); (Murphy and McKeogh, 2004)	Required (M); Less-intensive pollutant system is required (Murphy and McKeogh, 2004)	Required (L) [No separate pollutant control required, as it is integrated to Pollution control system of cement plant (Garcia et al., 2014)
	Sustainability Rank	3	4	2	1
O3	Retention/residence time /Process Time (Solid residence time)	300–3600 s (Singh et al., 2011)	60 s (Nixon, Dey, et al., 2013a)	1800s (Singh et al., 2011)	14 s (Baidya et al., 2017a)
	Sustainability Rank	3	2	3	1
O4	Overall system efficiency	28–42% (Hammond et al., 2011)	18–26% (Lombardi et al., 2015)	22–30% (Nixon, Dey, et al., 2013b; Yap and Nixon, 2015)	> 80% as the entire energy gets utilized except the system losses
	Sustainability Rank	2	4	3	1
O5	Volume reduction	50–90% (Singh et al., 2011)	80–90% (Singh et al., 2011)	80–95% (Arena, 2012)	100% (as no by-product is formed) (Baidya et al., 2017b)
	Sustainability Rank	4	3	3	1

(continued)

Table 3.11 (continued)

Factor Serial	Factors for Sustainability Index of Recovery	Waste Recycling by Pyrolysis	Waste Recycling by Incineration	Waste Recycling by Gasification	Waste Recycling by Co-processing
O6	Ash produced as residue	High (H); Organic portion produces ash as in incineration/WtE. But numerical value is available	High (H); 25 wt% of [90% bottom ash and ~10% fly ash]. (Amal et al., 2019)	High (H); Organic portion produces ash as in incineration/WtE. But numerical value is available	0% (WBCSD, 2014)
	Sustainability Rank	3	4	4	1
O7	Land requirements	0.8 hectare (300 tpd–plant) (Saini et al., 2012)	0.8 hectare (300 tpd–plant) (Lombardi et al., 2015)	0.8 hectare (300 tpd–plant) (Yap and Nixon, 2015)	0.6 hectare (300–400 tpd–pre-processing plant)
	Sustainability Rank	**3**	**4**	**3**	**1**
	SIR (Operational)	*22*	*22*	*22*	*09*
Economical Requirements for sustainability					
E1	Capital Cost for 100–200 ktpa	927 $/tpa (Yassin et al., 2009)	136–295 $/tpa (Nixon et al., 2013b; Yap and Nixon, 2015)	170 to 300 $/tpa (Yap and Nixon, 2015)	80–100 $/tpa
	Sustainability Rank	4	2	3	1
E2	Operational and Maintenance Cost for 100–200 ktpa	185 $/tonne (Nixon et al., 2014)	85 $/tonne (Chakraborty et al., 2013)	65–112 $/tonne (Murphy and McKeogh, 2004)	15–20 $/tonne
	Sustainability Rank	4	3	2	1
E3	Pre-treatment cost	Medium–High (segregation and shredding)	None (Nixon, Dey, et al., 2013a)	Medium–High (segregation and shredding) (Arena, 2012)	Medium (Mainly Segregation)
	Sustainability Rank	**4**	**1**	**4**	**2**

(continued)

Table 3.11 (continued)

Factor Serial	Factors for Sustainability Index of Recovery	Waste Recycling by Pyrolysis	Waste Recycling by Incineration	Waste Recycling by Gasification	Waste Recycling by Co-processing
	SIR (Economic)	*12*	*6*	*09*	*04*
	Environmental Requirements for sustainability				
En1	CO_2 emission	138 g CO_2/KWhe (Gaunt and Lehmann, 2008)	220 g CO_2/kWhe (Murphy and McKeogh, 2004)	114 g CO_2/kWhe (Nixon, Dey, et al., 2013a)	Negative Emission has been reported (Chatziaras et al., 2014)
	Sustainability Rank	3	4	2	1
En2	By-product generation by volume of input waste	30–35% (Hammond et al., 2011)	26–35% (Bosmans et al., 2013)	5–25% (Al-Salem et al., 2009)	0% (as the waste forms a part of the clinker) (Damtoft et al., 2008)
	Sustainability Rank	4	3	2	1
En3	Different waste streams which can be disposed of/utilized in a single setup	MSW (Pre-Treated) Industrial Waste (Selected streams)	MSW (Pre-Treated) Industrial Waste (All fragments of non-hazardous)	MSW (Pre-Treated) Industrial Waste (Selected streams)	MSW (Pre-Treated) Industrial Waste (All fragments hazardous and non-hazardous)
	Sustainability Rank	4	2	3	1
En4	By-product characteristics nature	Char, (Hornung et al., 2011)	Ash (Yap and Nixon, 2015)	Char, (Arena, 2012)	No by-product (Damtoft et al., 2008)
	Sustainability Rank	**3**	**4**	**3**	**1**
	SIR (Environmental)	*14*	*13*	*10*	*04*

Social Requirements for sustainability

(continued)

Table 3.11 (continued)

Factor Serial	Factors for Sustainability Index of Recovery	Waste Recycling by Pyrolysis	Waste Recycling by Incineration	Waste Recycling by Gasification	Waste Recycling by Co-processing
S1	Public Acceptance/Related issues	Accepted	Acceptance Issue exists Varies from country to country	Accepted	Highly Accepted [no new land required; done in existing cement plant; Emission issue is less]
	Sustainability Rank	3	4	3	1
S2	Odour	Medium	High	Medium	Low
	Sustainability Rank	3	4	2	1
S3	Perceived pollution issue	Medium	High	Medium	Low
	Sustainability Rank	3	4	2	1
S4	Noise Problem	Medium	High	Medium	Medium–High
	Sustainability Rank	2	4	1	3
S5	Expertise requirement	High (Higher skill)	Medium (Ordinary skill)	High (High skill)	High (Skill)
	Sustainability Rank	4	1	4	2
S6	Employment generation	Medium (Depends on size)	Medium (Depends on size)	Medium (Depends on size)	Low (Done in existing Cement plant)
	Sustainability Rank	2	2	2	1
	SIR (Sub Total)	**17**	**19**	**14**	**9**
	SIR (Total)	*65*	*60*	*55*	*26*

The sustainability index of recovery (SIR) may be defined as the sum of all the sustainability ranks of identified reasonable factors of sustainability of waste treatment processes under all the four pillars of sustainable development, such as operational, economic, environmental, and social requirements, while the lowest number of SIR represents the most sustainable process.

Considering seven factors (O1 to O7) in Operational Requirements for sustainability, three factors (E1 to E3) in Economical Requirements for sustainability, four factors (En1 to En4) in Environmental Requirements for sustainability, and six factors (S1 to S6) in Social Requirements for sustainability, the sustainability index of recovery (SIR) was calculated and the co-processing of wastes in cement kiln has been found to have the minimum sustainability index of recovery (SIR) number meaning the most sustainable waste management treatment process.

3.7 Scopes of Co-processing in the Indian Cement Industry

It has been observed from different research and ongoing operations in cement plants that there is a huge scope of utilization of different types of wastes in co-processing as AFR in cement plants. The types and possible sources have already been mentioned in Tables 3.1 and 3.2 and Fig. 3.1.

The Indian cement industry demands that these sectors of the industry operate in the most energy-efficient way in the world. In 2010, about 137 million $MtCO_2$ was emitted by the Indian cement industry which is equivalent to a 7% share of India's total man-made CO_2 emissions (Low carbon technology for Indian Cement Industry). Utilization of alternate fuels can reduce carbon emissions in the Indian cement industry. By the year 2025, the market demand for cement is estimated to reach 550–600 million tons per annum (MTPA) (Department of Commerce, Ministry of Commerce and Industry, Government of India). The projected growth and fuel requirements in Indian Cement industry by 2025 are presented in Table 3.12.

3.8 Conclusion

Co-processing is a preferred option for utilizing wastes. There are other energy recovery options where the amount of residue that is possible to generate varies from 15 to 25% while the co-processing is a zero-residue process of waste treatment. Using AFR in co-processing has several benefits. A few of those are mentioned below.

- Reduced the use of mined natural resources such as limestone, bauxite, and Iron ore and non-renewable fossil fuels such as coal.
- Conventional fuel reduction by Volume (20–25%) and TSR (10–15%).
- PAT target to reduce 5% MTOE will be helpful by using AFR.

Table 3.12 Projected growth and fuel requirements in Cement industry by 2025 in India. *Source* CII, (2016)

S.No	Parameters	Existing[3] 2010	Anticipated[4] 2025	Units
1	Cement production	217	600	Million TPA
2	Cement to clinker ratio	1.35	1.49	
3	Clinker production	161	402	Million TPA
4	Specific energy consumption	725	703	Kcal/kg of clinker
5	Total Thermal Energy Required	11.64	28.26	Million TOE
6	Average Energy from imported Coal	5500	5500	Kcal/kg of coal
7	Quantity of coal required	21.17	51.38	Million TPA
8	AF usage in TSR %	< 1	25	
9	Energy from alternate fuel estimated @ 25% of total energy		7.07	Million TOE

- Contributes to lowering emissions of GHG by replacing the use of fossil fuels. Reduction of global emission.
- High temperature around 1400 °C and long Calciner residence time around 5–6 s.
- Double valorization: organic and minerals totally destroyed.
- High efficiency and total recovery.
- Reduced land requirements for landfill option.
- No adverse effect in cement product quality.
- Reduces the burden on landfilling.
- Enhances business potential and reduces the cost of operating TSDF for hazardous wastes.
- Maximizes the recovery of resources from the waste.
- Employment generation and wider scopes of business.

The subsequent chapters will describe the aspects of cement manufacturing, pre-processing and co-processing, AFR, and other related issues.

References

Al-Salem, S. M., Lettieri, P., & Baeyens, J. (2009). Recycling and recovery routes of plastic solid waste (PSW): A review. *Waste Management, 29*(10), 2625–2643.

Arena, U. (2012). Process and technological aspects of municipal solid waste gasification. A review. *Waste Management, 32*(4), 625–639.

Amal, S., Al-Rahbi, & Paul T. Williams. (2019). Waste ashes as catalysts for the pyrolysis–catalytic steam reforming of biomass for hydrogen-rich gas production. *Journal of Material Cycles and Waste Management, 21,* 1224–1231. https://doi.org/10.1007/s10163-019-00876-8.

Ashraf, M. S., Ghouleh, Z., & Shao, Y. (2019). Production of eco-cement exclusively from municipal solid waste incineration residues. *Conservation & Recycling, 149*(2019), 332–342. https://doi.org/10.1016/j.resconrec.2019.06.018.

Angela J. Nagleb, Emma L. Delaney, Lawrence C. Bank, Paul G. Leahy. (2020). A comparative life cycle assessment between landfilling and co-processing of waste from decommissioned Irish wind turbine blades. *Journal of Cleaner Production, 277*, 123321. https://doi.org/10.1016/j.jclepro.2020.123321.

Bosmans, A., Vanderreydt, I., Geysen, D., et al. (2013). The crucial role of Waste-to-Energy technologies in enhanced landfill mining: a technology review. *Journal of Cleaner Production, 55*, 10–23.J.

Baidya, R., Ghosh, S. K., & Parlikar, U. V. (2016a). Co-processing of industrial waste in cement kiln–a robust system for material and energy recovery. *Procedia Environmental Sciences, 31*, 309–317.

Baidya, R., Ghosh, S. K., & Parlikar, U. V. (2016b). Co-processing of industrial waste in cement kiln—A robust system for material and energy recovery. *Procedia Environmental Sciences, 31*(2016), 309–317.

Baidya, R., Ghosh, S. K., & Parlikar, U. V. (2017a). Sustainability of cement kiln co-processing of wastes in India: A pilot study. *Environmental Technology, 38*(13–14), 1650–1659.

Baidya, R., Ghosh, S. K., & Parlikar, U. V. (2017b). *Environmental Technology, 38*(13–14), 1650–1659.

Baidya, R., Kumar Ghosh, S., & Parlikar, U. V. (2019). Blast furnace flue dust co-processing in cement kiln–A pilot study. *Waste Management & Research, 37*(3), 261–267.

Baidya, Rahul. (2019). Study of sustainable technology for energy recovery from waste, Ph.d. thesis 2019 at Jadavpur University, India.

Baidya, R., & Ghosh, S. K. (2019). Low carbon cement manufacturing in India by co-processing of alternative fuel and raw materials. *Energy Sources, Part a: Recovery, Utilization, and Environmental Effects, 41*(21), 2561–2572.

Baidya, R., & Ghosh, S. K. (2020). Co-processing of industrial trade rejects in cement plant. *Waste Management & Research, 38*(12), 1314–1320.

Chakraborty, M., Sharma, C., Pandey, J., et al. (2013). Assessment of energy generation potentials of MSW in Delhi under different technological options. *Energy Conversion and Management, 75*, 249–255.

Chatziaras, N., Psomopoulos, C. S., & Themelis, N. J. (2014). Use of alternative fuels in cement industry. In: *Proceedings of the 12th International Conference on Protection and Restoration of the Environment, ISBN*, 2014, pp. 978–960.

CII. (2016). *Promoting alternate fuel & raw material usage in Indian cement industry* (p. 2016). Supported by SHAKTI Sustainable Energy Foundation.

Damtoft, J. S., Lukasik, J., Herfort, D., et al. (2008). Sustainable development and climate change initiatives. *Cement and Concrete Research, 38*(2), 115–127.

Da-Hai Yan, Kare H. Karstensen, Qi-Fei Huang, Qi Wang, & Min-Lin Cai. (2010). Coprocessing of industrial and hazardous wastes in cement Kilns: A review of current status and future needs in China. *Environmental Engineering Science, 27*(1), 2010 DOI:10.1089=ees.2009.0144.

Dipl.-Ing. Sebastian Spaun et al. (2015). EinsatzalternativerRohstoffeimZementherstellungsprozess - Hintergrundwissen, technischeMöglichkeiten und Handlungsempfehlungen , Austrian Standard Önorm S 2100 http://www.zement.at/downloads/positivliste3.pdf.

Gaunt, J. L., & Lehmann, J. (2008). Energy balance and emissions associated with biochar sequestration and pyrolysis bioenergy production. *Environmental Science & Technology, 42*(11), 4152–4158.

Garcia, R. I., Moura, F. J., Bertolino, L. C., & de Albuquerque, B. E. (2014). Industrial experience with waste coprocessing and its effects on cement properties. *Environmental Progress & Sustainable Energy, 33*(3), 956–961.

Gunasegaram, D. R., & Molennar, D. (2015). Towards improved energy efficiency in the electrical connections of Hall-Héroult cellsthrough finite element analysis (FEA) modeling. *Journal of Cleaner Production, 93,* 174–192.

Guo, X., Shi, H., Wu, K., Ju, Z., & Dick, W. A. (2016). Performance and risk assessment of alinite cement-based materials from municipal solid waste incineration fly ash (MSWIFA). *Materials and Structures, 49*(6), 2383–2391. https://doi.org/10.1617/s11527-015-0655-x.

Ghosh, Sadhan K. (2017). State of the 3Rs in Asia and the Pacific, The Republic of India, Institute for Global Environmental Strategies (IGES), United Nations Centre for Regional Development (UNCRD), Japan.

Global Cement and Concrete Association (2018). GCCA Sustainability Guidelines for co-processing fuels and raw materials in cement manufacturing.

Gisele de Lorena Diniz Chaves, Renato Ribeiro Siman, & Ni-Bin Chang. (2021). Policy analysis for sustainable refuse-derived fuel production in Espírito Santo, Brazil. *Journal of Cleaner Production, 294,* 126344. https://doi.org/10.1016/j.jclepro.2021.126344.

Hanehara, S. (2001). Eco-cement and eco concrete environmentally compatible cement and concrete technology. *JCI/KCI International Joint Seminar, Kyonju, South Korea.* https://doi.org/10.1007/s10853-009-3342-x.

Hammond, J., Shackley, S., Sohi, S., et al. (2011). Prospective life cycle carbon abatement for pyrolysis biochar systems in the UK. *Energy Policy, 39*(5), 2646–2655.

Hornung, A., Apfelbacher, A., & Sagi, S. (2011). Intermediate pyrolysis: A sustainable biomass-to-energy concept-biothermal valorisation of biomass (BtVB) process.

https://doi.org/10.1016/S0921-3449(00)00077-X. S. Kerdsuwan, K. Laohalidanond, K., & Gupta Ashwani. (2020, February). Upgrading refuse-derived fuel properties from reclaimed landfill using torrefaction. *Journal of Energy Resources; Technology, 143.* DOI: https://doi.org/10.1115/1.4047979.

https://doi.org/10.1080/09593330.2020.1856191; https://www.cement.org/docs/default-source/market-economics-pdfs/cementindustry-by-state/usa-statefacsheet-17-d2.pdf?sfvrsn=e77fe6 bf_2. in portland cement clinker. Chem. Eng. Res. Des. 21, 757–762. https://doi.org/10.

Ismael Vemdrame Flores, Felipe Fraiz, Rafael Adriano Lopes Junior, Maurício Covcevich Bagatini. (2019). Evaluation of spent pot lining (SPL) as an alternative carbonaceous material in iron making processes, j m a t e r r e s t e c h n o l . 2 0 1 9;8(1), 33–40.

John R. Fyffe, Alex C. Breckel, Aaron K. Townsend, & Michael E. Webber. (2016). Use of MRF residue as alternative fuel in cement production, Waste Management.

Jolanta Sobik-Szołtysek, Katarzyna Wystalska. (2019). Coprocessing of sewage sludge in cement kiln, Editor(s): Majeti Narasimha Vara Prasad, Paulo Jorge de Campos Favas, Meththika Vithanage, S. Venkata Mohan, Industrial and Municipal Sludge, Butterworth-Heinemann, pp. 361–381, ISBN 9780128159071, https://doi.org/10.1016/B978-0-12-815907-1.00016-7.

Kyle A. Claviera, Benjamin Wattsb, C, Yalan Liua, Christopher C. Ferraroc, & Timothy G. Townsenda. (2019). Risk and performance assessment of cement made using municipal solid waste incinerator bottom ash as a cement kiln feed. *Resources, Conservation & Recycling, 146*(2019), 270–279. https://doi.org/10.1016/j.resconrec.2019.03.047.

Lam, C. H. K., Barford, J. P., & Mckay, G. (2010). Utilization of incineration waste ash residues.

Lam, C. H. K., Barford, J. P., & McKay, G. (2011). Utilization of municipal solid waste incineration ash in Portland cement clinker. *Clean Technologies and Environmental Policy 13*(4)

LI Chun-ping. (2012). Pilot experiments on co-processing oversized products screened from aged refuse in cement kiln. *Advanced Materials Research, 518–523,* 3421–3426. doi:https://doi.org/10.4028/www.scientific.net/AMR.518-523.3421.

Lombardi, L., Carnevale, E., & Corti, A. (2015). A review of technologies and performances of thermal treatment systems for energy recovery from waste. *Waste Management, 37,* 26–44.

Low carbon technology for Indian Cement Industry.

Millano, E. F. (1996). Hazardous waste: Storage, disposal, remediation, and closure. *Water Environment Research, 68*(4), 586–608.

Murphy, J. D., & McKeogh, E. (2004). Technical, economic and environmental analysis of energy production from municipal solid waste. *Renewable Energy, 29*(7), 1043–1057.

Michaeleliasboesch, Annettekoehler, & Stefaniehellweg. (2009). Model for cradle-to-gate life cycle assessment of clinker production. *Environmental Science & Technology, 43,* 7578–7583.

Moses P. M. Chinyama. (2011). Alternative fuels in cement manufacturing, alternative fuel. In Dr. Maximino Manzanera (Ed.), ISBN: 978-953-307-372-9. InTech, Available from: http://www.int echopen.com/books/alternative-fuel/alternative-fuels-in-cement-manufacturing.

Madlool, N. A., Saidur, R., Rahim, N. A., & Kamalisarvestani, M. (2013). An overview of energy savings measures for cement industries. *Renew Sust Energ Rev, 19*, 18–29.

Ministry of Housing and Urban Affairs (2018). Guidelines on Usage of Refuse Derived Fuel in Various Industries.

Mohamed M. Elfaham, & Usama Eldemerdash. (2018). Advanced analyses of solid waste raw materials from cement plant using dual spectroscopy techniques towards co-processing. *Optics and Laser Technology, 111*, 338–346. https://doi.org/10.1016/j.optlastec.2018.10.009.

Michael Hinkel, Daniel Hinchliffe, Dieter Mutz, Steffen Blume, & Dirk Hengevoss. (2019). Guidelines on pre- and co-processing of waste in cement production. Deutsche Gesellschaft für, InternationaleZusammenarbeit, GmbH (GIZ).

Nema, A. K., & Gupta, S. K. (1999). Optimization of regional hazardous waste management systems: An improved formulation. *Waste Management, 19*(7–8), 441–451.

Nixon, J. D., Dey, P. K., Ghosh, S. K., et al. (2013a). Evaluation of options for energy recovery from municipal solid waste in India using the hierarchical analytical network process. *Energy, 59*, 215–223.

Nixon, J. D., Wright, D. G., Dey, P. K., et al. (2013b). A comparative assessment of waste incinerators in the UK. *Waste Management, 33*(11), 2234–2244.

Nixon, J. D., Dey, P. K., Davies, P. A., et al. (2014). Supply chain optimisation of pyrolysis plant deployment using goal programming. *Energy, 68*, 262–271.

Pan, J. R., Huang, C., Kuo, J. J., & Lin, S. H. (2008). Recycling MSWI bottom and fly ash as raw materials for Portland cement. *Waste Management, 28*(7), 1113–1118. https://doi.org/10.1016/j. wasman.2007.04.009.

Planning Commission. (2014). Report by the Planning Commission "Task Force on Waste to Energy," India.

Parlikar, U., Bundela, P. S., Baidya, R., Ghosh, S. K., & Ghosh, S. K. (2016). Effect of variation in the chemical constituents of wastes on the co-processing performance of the cement kilns. *Procedia Environmental Sciences, 35*, 506–512.

Press Information Bureau. (2016). Solid Waste Management Rules Revised 25. after 16 years; Rules now extend to Urban and Industrial Areas: Javadekar". Ministry of Environment, Forest and Climate Change, April 5, 2016.

Peter, C. H, Martin, L. (2019). Lea's chemistry of cement and concrete. 5th ed: Butterworth-Heinemann: Academic Press, pp. 31–56.

Rigo & Rigo Associates. (1995). An analysis of technical issues pertaining to the determination of MACT standards for the waste recycling segment of the cement industry. Environmental Risk SciencesInc., Environomics, and CKRC (Cement Kiln Recycling Coalition).

Ryunosuke Kikuchi. (2001). Recycling of municipal solid waste for cement production: Pilot-scale test for transforming incineration ash of solid waste into cement clinker.*Resources, Conservation and Recycling, 31*(2), 137–147, ISSN 0921-3449.

Richard Bolwerk. (2004). Co-processing of waste and Energy efficiency by cement plants, not mentioned.

Reza, B., Soltani, A., Ruparathna, R., Sadiq, R., & Hewage, K. (2013). Environmental and economic aspects of production and utilization of RDF as alternative fuel in cement plants: A case study of Metro Vancouver Waste Management. *Resources, Conservation and Recycling, 81*(2013), 105–114.

Rahman, A., Rasul, M. G., Khan, M. M. K., & Sharma, S. (2015). Recent development on the uses of alternative fuels in cement manufacturing process. *Fuel, 145*, 84–99.

Saikia, N., Kato, S., & Kojima, T. (2007). Production of cement clinkers from municipal solid waste incineration (MSWI) fly ash. *Waste Management, 27*(9), 1178–1189. https://doi.org/10.1016/j.wasman.2006.06.004.

Singh, R. P., Tyagi, V. V., Allen, T., et al. (2011). An overview for exploring the possibilities of energy generation from municipal solid waste (MSW) in Indian scenario. *Renewable and Sustainable Energy Reviews, 15*(9), 4797–4808.

Saini, S., Rao, P., & Patil, Y. (2012). City based analysis of MSW to energy generation in India, calculation of state-wise potential and tariff comparison with EU. *Procedia-Social and Behavioral Sciences, 37*, 407–416.

Samolada, M. C., & Zabaniotou, A. A. (2014). Comparative assessment of municipal sewage sludge incineration, gasification and pyrolysis for a sustainable sludge-to-energy management in Greece. *Waste Management, 34*(2), 411–420.

Trezza, M. A., & Scian, A. N. (2000). Burning wastes as an industrial resource: Their effect on Portland cement clinker. *Cement and Concrete Research, 30*(1), ISSN 0008–8846.

Volume 47, Part B, 2016, Pages 276–284, ISSN 0956–053X, https://doi.org/10.1016/j.wasman.2015.05.038.

Wang, L., Li, R., Li, Y., & Wei, L. (2012). Incorporation of cadmium into clinker during the co-processing of waste with cement kiln. *Advanced Materials Research, 347–353*(2012), 2160–2164.

Wendell de Queiroz Lamas, Jose Carlos Fortes Palau, Jose Rubens de Camargo. (2012). Waste materials co-processing in cement industry: Ecological efficiency of waste reuse. *Renewable and Sustainable Energy Reviews, 19*(2013), 200–207. http://dx.doi.org/https://doi.org/10.1016/j.rser.2012.11.015.

WBCSD. (2014). Guidelines for Co-Processing Fuels and Raw Materials in Cement Manufacturing.

Yassin, L., Lettieri, P., Simons, S. J., et al. (2009). Techno-economic performance of energy-from-waste fluidized bed combustion and gasification processes in the UK context. *Chemical Engineering Journal, 146*(3), 315–327.

Yufei Yang, Yu Yang, Qunhui Wang, & Qifei Huang. (2011). Release of heavy metals from concrete made with cement from cement Kiln co-processing of hazardous wastes in pavement scenarios. *Environmental Engineering Science, 28*(1), 2011. DOI: https://doi.org/10.1089/ees.2010.0066.

Yap, H. Y., & Nixon, J. D. (2015). A multi-criteria analysis of options for energy recovery from municipal solid waste in India and the UK. *Waste Management, 46*, 265–277.

Yao, J., Kong, Q., Qiu, Z., Chen, L., & Shen, D. (2019). Patterns of heavy metal 655 immobilization by MSW during the landfill process. *Chemical Engineering Journal, 375*(122060), 656. https://doi.org/10.1016/j.cej.2019.12206.

Yeqing, Li. et al. (2020). Research on dioxins suppression mechanisms during MSW co-processing in cement kilns. *Procedia Environmental Sciences, 16*(2012), 633–640. doi: https://doi.org/10.1016/j.proenv.2012.10.087.

Ziegler, D., Schimpf, W., Dubach, B., Degré, J.-P., & Mutz, D. (2006). Guidelines on co-processing waste materials in cement production. The GTZ-Holcim Public Private Partnership.

Part III
Cement: Theory, Production Technology and Operations

Chapter 4
Cement Manufacturing—Technology, Practice, and Development

4.1 Introduction

The history of cement goes back to Roman Empire. The credit of the invention of the Portland cement was first produced by a British stone mason, Joseph Aspdin of Leeds, Yorkshire, England. A patent was approved to him for a material produced from a synthetic mixture of limestone and clay in 1824 (P. E. Halstead, 1961). A mixture of limestone and clay powder was taken in his kitchen. The mixture is then ground into a powder that creates cement in ball mill-like equipment. The powder is hardened when mixed with water. The name Portland was given by the inventor as it resembles a stone quarried on the Isle of Portland. The first use of modern-day Portland cement was in the tunnel construction in the Thames River in 1828. During the twentieth century, cement manufacture spread worldwide. By 2019, India and China have become the world leaders in cement production, followed by Vietnam, the United States, and Egypt.

The primary cement manufacturing process involves the mining of raw materials, mainly limestone and clay, which are used in cement manufacturing. In most of the cases, the limestone and clay are excavated from open cast mines by drilling and blasting and other appropriate processes in mines. Subsequently, the limestone and clay are loaded onto dumpers which transport the materials. Limestone is unloaded into hoppers of limestone crushers and the clay is unloaded into open yard storage. Then it is transported by trucks and unloaded into the hopper of a clay crusher. There are three types of clay used in cement manufacturing, namely silty clay, Zafarana clay, and Kaolin (Christopher Hall, 1976). The cement manufacturing process description illustrated here is derived from various literature sources. A comprehensive overview of cement manufacturing and technologies can be found in (CEMBUREAU, 1999; Duda, 1985; Environment Agency, 2001; IPPC, 2013; Karstensen, 2006).

© The Author(s), under exclusive license to Springer Nature Singapore Pte Ltd. 2022   73
S. K. Ghosh et al., *Sustainable Management of Wastes Through Co-processing*,
https://doi.org/10.1007/978-981-16-6073-3_4

4.2 Main Process Description

There are four main types of processes used in cement manufacture: Dry Process, Semi-dry Process, Semi-wet Process, and the Wet process. In all these processes, the following steps are involved:

- Quarrying.
- Raw materials preparation.
- Fuel Preparation.
- Clinker Manufacturing Process.
- Cement grinding.
- Cement dispatch.

All these steps are described in detail in the next sections.

4.2.1 Quarrying

The major raw material used in cement manufacture is lime. It is derived from materials such as limestone, chalk, marl, shale, and clay, which are obtained from mines by quarrying them. These mines are generally located near the cement plant. These materials are then crushed and transported to the cement plant where they are stored, homogenized, blended with corrective materials, and processed further for cement manufacture. Figure 4.1 demonstrates the limestone quarry in Michigan, China, and India.

"Corrective" materials utilized are bauxite, iron ore, or sand that supplement the Aluminium, iron, and silica requirement. The corrective materials may also be other materials such as laterite, red occur, and clay. The quantities of these corrective materials are usually very small in few percentages and depend upon the raw mix design. All the raw materials and corrective materials are mined materials using quarrying techniques. To some extent, the corrective materials can be replaced with secondary or alternative raw materials as well. These alternative materials get produced from industrial processes; for example, red mud from the Aluminium industry, mill scale from rolling mills, rice husk ash from rice husk-fired boilers, etc.

4.2.2 Raw Materials Preparation

In Dry process and Semi-dry process plants, all the raw materials are mixed in desired proportions to produce a raw meal and are ground to the required size. In the wet and semi-wet process, the raw materials are mixed and made into a slurry. This slurry is then ground to the defined size. The raw meal or the raw material slurry is finely ground and homogenized in raw meal silos or basins. This helps to achieve the

Fig. 4.1 a. Largest Limestone Quarry in Michigan. **b** Limestone quarry in China. **c** Limestone Quarry in Gujarat, India (Formed by calcite, calcium carbonate, CaCO3)

uniform chemical composition required for manufacturing desired quality of cement. To produce one ton of clinker, approximately 1.5 tons of raw materials are required.

A typical flow sheet of the modern dry process cement plant is presented in Fig. 4.2.

4.2.3 Fuel Preparation

The most common type of fossil fuel used in the cement industry is coal. "Alternative" fuels that are derived from wastes generated in industrial, municipal, and agricultural sectors are also widely used today as fuels in cement manufacture. Fossil fuels preparation consists of crushing, drying, grinding, and homogenizing which takes place at the plant site. Specific installations are required for fuel preparation such as grinding mills and silos for solid fuels and tanks for liquid fuels. The basic process design applied in the clinker manufacturing process largely governs the thermal fuel consumption.

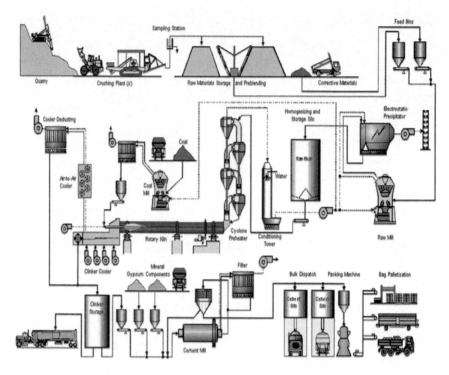

Fig. 4.2 Modern dry process cement production process

4.2.4   Manufacturing of Cement Clinker

The prepared raw material is called the "kiln feed" and is fed to the kiln. In the kiln, this kiln feed is subjected to a thermal treatment process which consists of drying, preheating, calcination, and clinkerization. The max temperature reached by the material in the kiln during clinkerization is around 1450 °C. The product, i.e., clinker is then cooled down with air to below 200 °C and is then stored prior to final grinding.

Figure 4.3 depicts the flow sheet of a Rotary kiln process with cyclone preheater and gas dust collection system.

The kiln system in which the clinkerization takes place is a rotary kiln with or without additional infrastructure called as "suspension preheaters" and pre-calciner. The rotary kiln is a rotating steel tube with a length to diameter ratio between 10 and 40. A slight inclination of 2.5 to 4.5 degrees is provided for a smooth flow of material. The kiln normally rotates at a speed of 0.5–4.5 rounds-per-minute (RPM).

Exhaust heat from the kiln is used to dry raw meal, fuels, or other additives utilized in the cement manufacture. Exhaust gases are then let out after dedusting the same in ESPs or Bag filters.

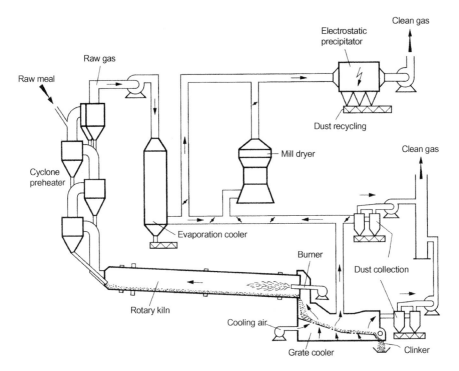

Fig. 4.3 Rotary kiln with cyclone preheater and gas dust collection

4.2.5 Cement Grinding

There are two main types of cement, ordinary Portland cement (OPC) and various blended cements. OPC is produced by grinding clinker with a few percent of gypsum in a cement mill; The blended one is produced by the addition of pozzolanic materials to OPC. These pozzolanic materials include granulated blast-furnace slag, volcanic ash, fly ash, or inert fillers such as limestone. These blended cements could be of different kinds such as Portland pozzolanic cement (PPC) or Portland Slag cement (PSC) or Composite Cement. The extent of the addition of these pozzolanic or inert materials to the clinker depends upon their physico-chemical characteristics and also the permitted norms by the authorities. The clinker content in the blended cement is called as the clinker factor. The different cement types need to be put in separate storage silos before they are sent out in the market for selling.

The grinding operation of the cement is carried out in cement mills located in the grinding plants. The grinding plants can be installed at a far away location from the clinkering plant.

4.2.6 Cement Dispatch

Cement is shipped in bulkers or in packed bags. Transport can be by road, railway, and waterways and depends on local conditions.

4.3 Characteristics of Material Used in Cement Production

Clinker production requires a mixture of calcium, silicon, aluminium, and iron as the main elements. Appropriate clinker phases get formed when they are mixed in the correct proportions and subjecting them to the clinkerization temperature of 1450 °C. This temperature is achieved by firing the fuel in the kiln in the main burner and in the pre-calciner.

Figure 4.4 exhibits the main burner mounted at the kiln outlet.

There are four steps in clinker formation:

i.   Drying and preheating of the raw meal take place while heating from room temperature to 900 °C. During this phase, free and chemically bound water gets released.
ii. Calcination reaction takes place from 600–900 °C during which CO_2 from limestone gets released.
iii. Clinkerization reactions take place from 1250–1450 °C during which initiation of the formation of crystalline phases takes place.

Fig. 4.4 Main burner at the kiln outlet

iv.   While cooling the liquid phase from 1350–1200 °C, the formation of the crystalline phases takes phase.

Mineral composition and its structure in the clinker determine the properties of the clinker. Alkalis, sulphur, and chlorides present in the kiln get volatilized at high temperatures and remain as circulating elements by condensation. These elements leave the kiln with the clinker. If these circulating elements are high in quantum, then it becomes necessary to instal a "bypass" system where a part of the dust laden exhaust gases of the rotary kiln are extracted out from the system.

4.3.1 Raw Mix Constituents

Raw mix in clinker manufacturing consists of naturally occurring mined materials as main materials. These consist of materials that are rich in Calcium, Aluminium, Iron, and Silica. Small quantities of correctives are also used in the cement manufacture, needed to adjust the chemical composition to the required quality standards.

A raw mix is prepared based on the raw materials chemistry, process design, and process considerations, but will also be dependent upon product specifications and on environmental considerations. To achieve good quality clinker, a well-designed raw mix, adequately ground raw meal, and uniform composition are essential requirements together with a smooth kiln operation. Hence, quality monitoring and control of the inputs, process, product, and emissions through appropriate sampling and evaluations are important considerations in cement manufacture.

4.3.2 Usage of Fuels

Petcoke, coal, heavy oil, and natural gas are the main fossil fuels used in the manufacturing of cement. Petcoke and coal are the major ones. Tyres or Tyre-Derived Fuels, waste oil, plastics or refuse-derived fuels (RDF), solvents, and many other industrial waste streams are used as "alternative" fuels. As in the case of fossil fuels, the composition of the ash from AFRs also needs to be considered in the raw mix design because it combines with the raw materials and will be incorporated in the clinker product.

A certain amount of non-volatile heavy metals will be present in the fossil or the alternative fuels and will be incorporated into the clinker structure as a non-leachable constituent. Excessive amounts will be loosely bound and may be released as leachable metals. Hence, it is desired to monitor and control these heavy metals in the input streams. Other heavy metals such as mercury, thallium, or cadmium may be volatilized and be released out of the stack or captured in the dust. Therefore, these also need to be monitored and controlled.

4.4   Different Cement Manufacturing Processes

The manufacturing of cement clinker was initially based on the "wet" process, but the "dry" process is today the state of the art. The "semi-wet" and "semi-dry" processes were intermediate.

4.4.1   The Dry Process

The raw meal is prepared by drying and grinding in ball mills or vertical roller mills. Drying is achieved by using the hot gases or air from the kiln or the cooler. Homogenizing is done in silo or dome systems. The dried raw material is fed to the top cyclone of the preheater system in preheater kilns and the raw meal moves down the preheater system in a step-wise manner while having counter-current heat exchange with the hot exhaust gases from the rotary kiln that are moving counter-currently up the preheater system. The preheater cyclones provide excellent contact and heat exchange between raw meal and hot gas. The cyclones help in separating solids and gases at each stage. The exhaust gases leave the preheater at a temperature of 300–360 °C and this gas is generally utilized for raw material drying.

Figure 4.5 exhibits the present-day modern dry process kiln system with preheater and pre-calciner.

Fig. 4.5 Modern dry preheater process with pre-calciner

The preheated raw meal enters the calciner and then gets calcined in the pre-calciner. The calcined meal then enters the rotary kiln where it undergoes the required reactions at high temperature and forms a clinker.

In case excess alkalis or chlorine is present in the kiln system, then a bypass system needs to be installed. A bypass system will divert up to 15% of the kiln gas out of the kiln system and extract particulates enriched with alkalis or chlorine. The bypass duct is located at the kiln inlet and the lower part of the preheater. The gas is cooled with air and then released to the stack after filtration in another bag filter or ESP.

State-of-the-art suspension preheater kilns have 5–6 cyclone stages with a maximum capacity limited to approximately 10,000–12,000 ton clinker/day. Figure 4.6 depicts the dry process plant system with suspension preheater cyclone tower.

Fig. 4.6 Suspension preheater cyclone tower

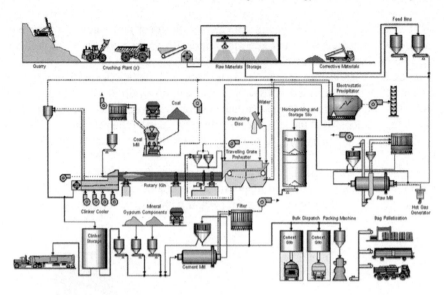

Fig. 4.7 Semi-dry process

4.4.2 The Semi-Dry Process

In this process, the raw meal is granulated and sent on an inclined rotating table, dropped onto a travelling grate preheater, and into the rotary kiln. The granulated material then gets dried, preheated, and calcined on the travelling grate. Figure 4.7 depicts the semi-dry process.

Due to the low temperatures of the kiln gases, they cannot be used for the raw material drying purpose.

4.4.3 The Semi-Wet Process

In the semi-wet process, the dewatered slurry from a filter press is used as the feed material. Typically, these filter cakes will have a residual moisture content of 16–21%. This process makes full use of the hot kiln gases and cooler air. Figure 4.8 depicts the semi-wet process.

4.4.4 The Wet Process

Wet process kilns are the old generation kilns used to produce clinker. The feed to this kiln is slurry which typically contains more than 40% of water. Homogenization

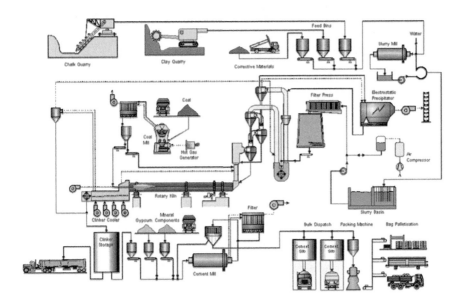

Fig. 4.8 Semi-wet process

is achieved in slurry basins where the slurry is continuously stirred using compressed air. The homogenized slurry is then pumped into the rotary kiln.

The water is evaporated in the drying zone of the rotary kiln and the heat exchange between the kiln feed and the combustion gases happens in the drying zone of the kiln, usually equipped with chains to facilitate drying. Conventional wet kilns have high heat consumption and produce large volumes of combustion gases containing water vapour. Figure 4.9 depicts the typical wet process.

4.4.5 Vertical Shaft Kiln

Vertical shaft kiln (VSK) is a low-volume and "obsolete" technology for cement manufacturing. VSK consist of a refractory-lined, vertical cylinder that is < 3 m in diameter and about 10 m tall. Raw meal pellets and fine-grained coal, called black meal, are mixed and fed from the top of VSK to produce clinker at the bottom. Shaft kilns produce less than 300 tonnes/day of clinker. Figure 4.10 provides a view of the VSK plant.

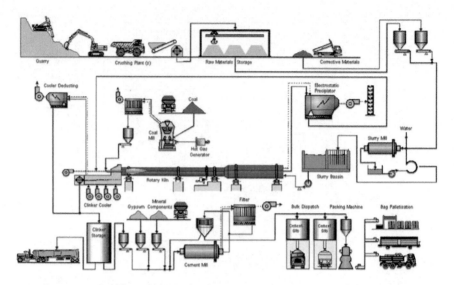

Fig. 4.9 Wet process of the cement production process

Fig. 4.10 Vertical shaft kiln cement production process

4.5 Kiln Exhaust Gases

The kiln exhaust gases from the manufacturing processes described above are treated for dust removal in an appropriate pollution control device. ESP and Bag filters are the two types of pollution control devices that are commonly used in the cement industry.

Bag filters use a synthetic high-temperature-resistant fabric to filter the dust. Exit gases pass through the bags while dust particles are captured on the surface; "reverse gas" and "pulse jet" filters are two types of bag filters used in these devices. The filter performance is not susceptible to process disturbances or "CO peaks". Electrostatic precipitators are using electric fields to separate dust from the gas, both coarse and finer particles. ESPs are susceptible to process changes such as CO peaks. The dedusting efficiency can be increased by making use of more than one electric "field" operating in series.

Figure 4.11 provides a view of a typical exhaust gas dedusting system implemented at the cement plant.

The dust collected by the pollution control devices is normally returned to the process by sending it to the raw materials preparation system. It can be also added to the cement mill if authorities or the standards permit the same. Dust control systems are necessary in the clinker cooler, the raw mill, and the cement mills.

The clinker leaving the kiln has a temperature around 1200 °C and needs to be cooled rapidly in appropriate clinker coolers. In this cooling process, heat from the clinker is picked by the cooling air and sent to the main kiln burner as secondary air,

Fig. 4.11 Exit gas dedusting system with the electrostatic precipitator (ESP)

Fig. 4.12 Kiln with planetary cooler

eventually to the pre-calciner as tertiary air. The following are the three main types of clinker coolers.

- Rotary cooler,
- Planetary cooler, and
- Grate cooler.

Figure 4.12 provides a view of the kiln system on which a planetary cooler is installed.

In a planetary cooler, several tubes are installed peripherally at the discharge end of the kiln. Hot clinker exchanges its heat with the incoming air coming from the tubes. Comparatively high wear and thermal shock effects are the drawbacks of the planetary coolers.

Grate coolers are the most preferred coolers in the current cement plant installations. The clinker layer travels slowly on a moving grate of perforated plates. The whole cooling zone includes two zones; this preheated air from the recuperation zone is sent as combustion air to the main burner and the pre-calciner. The hot air from the aftercooling zone is used for drying raw materials or coal. These coolers are widely used due to their high energy efficiency and flexible heat recovery.

4.6 Fuel Processing

The fuels require proper processing for their efficient use. The process depends upon the characteristics of the fuel. These fuels can be solid, liquid, or gaseous. The fuels can be conventional fossil fuels or alternative fuels from industrial, municipal, or agricultural sectors. The design of their storage, handling, preparation, and firing systems needs to be aligned to their properties. The most important aspect is that the fuel input has to be uniform, and its metering has to be reliable. Processing of coal and petcoke is done by grinding them to a specific fineness in ball mills, vertical roller mills, or impact mills. The entire fuel preparation system must be designed with appropriate fire protection measures. The ground coal or petcoke powder is stored in suitable storage systems and fed to the burner with appropriate metering and feeding systems. Fuel oil is stored in large tanks on site. These are provided with heating systems to raise their temperature of about 80 °C. For facilitating the smooth flow of the oil in the burner, its viscosity is reduced by heating the oil up to a temperature of above 120 °C. Figure 4.13 depicts the view of a gas-fired cement kiln.

Alternative fuels require specific treatment which is called pre-processing and they are fed to the kiln using an appropriate feeding system designed for the same. Alternative fuels are pre-processed in a facility that can be inside or outside the cement plant. Alternative fuel plants are often designed to handle different wastes originating from industrial, municipal, and agricultural sectors.

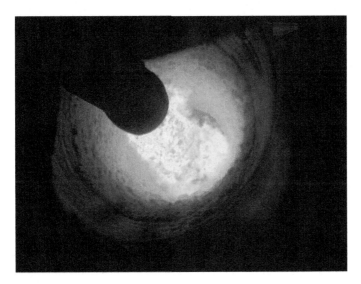

Fig. 4.13 Gas-fired cement kiln

4.7   Preparation of Pozzolanic and Mineral Materials

The addition of Pozzolanic and Mineral materials used in the manufacture of blended cements requires installations for processing them. The processing steps consist of storage, crushing, blending, and drying. Subsequently, it is used for feeding to the cement mill. Commonly used materials include volcanic rocks, limestone or calcined clay, granulated blast-furnace slag, pulverized fly ash from power stations, micro silica, etc.

Materials containing high moisture require drying. For this, various equipment such as flash driers and rotary tube driers are made use of. The heat for drying is derived from kiln gases or cooler air or with a hot gas source. Mineral materials may be inter-ground with cement clinker and gypsum in a cement mill or may be ground separately and blended with Portland cement subsequently.

4.8   Specific Features of Cement Production Process

For the cement-related reactions to occur, material temperatures of up to 1450 °C are required. For this, it is necessary to maintain peak combustion temperatures of about 2000 °C in the kiln with the main burner flame. The combustion gases from the main burner remain at a temperature above 1100 °C for 5–10 s. Excess oxygen at a level of 2–3% over stoichiometry is maintained in the combustion gases of the rotary kiln for the clinker to get burned appropriately under oxidizing conditions. These conditions help in the formation of the desired clinker phases and the quality of the finished cement.

The minimum retention time of the material in the rotary kiln would be 20 min and the maximum would be 60 min depending on the length of the kiln. The temperature profiles for the material and combustion gases in a preheater / pre-calciner kiln are illustrated in Fig. 4.6. The burning conditions in kilns with pre-calciner firing depend on the pre-calciner design. Usually, pre-calciner gas temperatures are around 950 °C and the retention time of the gas in the pre-calciner is >3 s. Material temperatures of up to 1450 °C, flame temperatures of up to 2000 °C, and gas retention times of up to 10 s at temperatures between 1100 and 2000 °C in the kiln are the thermal conditions prevailing in the kiln system. Organic materials fed to the main burner will be completely oxidized due to the high temperatures.

The cement manufacturing process operates continuously. Kiln operation is desired to be smooth to meet production and quality standards. To achieve these objectives, all process parameters are continuously monitored, recorded, and controlled. Further, analytical monitoring and control of all raw materials, fuels, intermediate, and finished products as well as environmental monitoring are carried out. Figure 4.14 illustrates the gas and material profiles inside the kiln.

A cement manufacturing process is well suited for co-processing of wastes from different sources. Wastes need to be converted into alternative fuels and raw materials

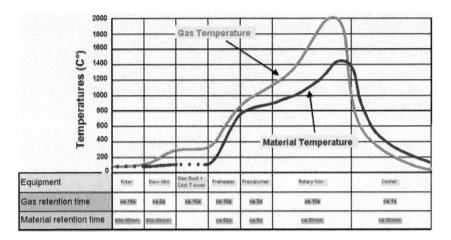

Fig. 4.14 Gas and material temperature profiles inside the kiln

(AFRs) through the pre-processing operation and then utilized to substitute fossil fuels and raw materials used in the cement process. For environmentally sound co-processing of AFRs, it is necessary to select the most appropriate feed point in the kiln system.

Materials that do not contain organic materials can be added to the raw meal or the raw slurry preparation system. Wastes containing significant quantities of organic materials should be introduced through the main burner or the pre-calciner. Pozzolanic and other minerals are fed to the cement mill.

Being a process in which a large quantum of materials get processed and produced in a continuous mode at high-temperature gas retention times, cement kilns can have a large capability of utilizing a variety of materials. Cement plants exercise several voluntary controls that are beyond the statutory provisions such as waste management happens in an appropriate manner in the cement kiln, sustain the required product quality, ensure the protection of the cement process from operational concerns, avoid negative impacts to the environment, and ensure the occupational health and safety of the operating personnel.

4.9 Cement Production in Developing Countries

The global cement industry is mostly expanding in emerging markets which are located in developing countries. When the companies build new plants in any country, usually they adopt the best available techniques (BAT) so as to be competitive and viable. Usually, in developing countries, the bigger cement companies tend to impose their internal standards on business ethics, labour rights, corporate responsibility, health, safety, and environment that are beyond the local statutory obligations.

References

CEMBUREAU. (1999). Best available techniques for the cement industry. The European Cement Association. Rue d'Arlon 55 - B-1040 Brussels. http://www.cembureau.be.

Christopher Hall. (1976). On the history of Portland cement after 150 years.*Journal of Chemical Education, 53*(4), 222. Publication Date: April 1, 1976, https://doi.org/10.1021/ed053p222.

Duda, W. H. (1985). *Cement data book*. Bauverlag Gmbh.

Environment Agency. (2001). Integrated pollution prevention and control—Guidance for the Cement and Lime sector. Environment Agency, SEPA and Environment and Heritage Service, Bristol, UK, April 2001. Retrieved from http://www.environment-agency.gov.uk/.

HALSTEAD, P. E. (2014). The early history of Portland cement, transactions of the Newcomen Society, Vol 34, 1961; Issue 1, Pages 37–54 | Published online: 31 Jan 2014; https://doi.org/10.1179/tns.1961.003.

IPPC. (2013). Best Available Techniques (BAT) Reference Document for the Production of Cement, Lime and Magnesium Oxide: Industrial Emissions Directive 2010/75/EU. 978-92-79-32944-9.

Karstensen, K. H. (2006). Formation and Release of POPs in the Global Cement Industry—Second Edition. Report to the World Business Council for Sustainable Development. 30 January. Retrieved from http://www.wbcsdcement.org/pdf/formation_release_pops_second_edition.pdf http://www.wbcsdcement.org/pdf/formation_release_pops_second_edition.pdf.

Chapter 5
Fundamentals of Cement Chemistry, Operations, and Quality Control

5.1 Introduction

Co-processing involves utilizing waste materials as Alternative Fuels and Raw materials (AFRs). Their role in cement kilns, therefore, represents the same importance as the one that is applicable to natural fuels, namely coal, other fossil fuels, and raw materials. Whatever technical and scientific considerations are extended to the natural raw materials and fuels used in cement manufacturing, the same importance needs to be extended to AFRs as well. The relevance of these principles becomes more important with the increased quantum of usage of AFRs in the cement kilns. This chapter provides relevant understanding and inputs on different aspects related to subjects such as cement chemistry, laboratory evaluations, thermal energy, and kiln operational aspects. (CII-GBC, 2010).

Raw Materials necessary for cement manufacture include materials containing Calcium, Silica, Alumina, and Iron. These raw materials are mixed in certain proportions and burnt at 1400 to 1450 °C in a rotary kiln to form a clinker which is subsequently ground to powder and mixed with 2 to 8% gypsum. This is called as ordinary Portland cement (OPC). OPC is manufactured in three grades; 33 Grade, 43 Grade, and 53 Grade.

A 33-grade cement concrete mortar made with cement, water, and sand in the ratio of 1:1:3 cured for 28 days under controlled conditions achieves a minimum compressive strength of 33 N/mm². A 33-grade cement is used in plain cement concrete (PCC) where it is not subjected to stress. It complies with the specifications set out in BIS 269 standards. (BIS 269, 2013).

The 43-grade cement concrete mortar made with cement, water, and sand in the ratio of 1:1:3,cured for 28 days under controlled conditions achieves a minimum compressive strength of 43 N/mm². A 43-grade cement is used in general civil engineering construction work, RCC structures (M25 concrete), Non-RCC structures, plastering works, etc. It complies with the specifications set out in BIS IS 8112 standards. (BIS 12269, 2013).

© The Author(s), under exclusive license to Springer Nature Singapore Pte Ltd. 2022   91
S. K. Ghosh et al., *Sustainable Management of Wastes Through Co-processing*,
https://doi.org/10.1007/978-981-16-6073-3_5

The 53-grade cement concrete mortar made with cement, water, and sand in the ratio of 1:1:3 cured for 28 days under controlled conditions achieves a minimum compressive strength of 53 N/mm^2. The 53-grade cement is used where early strength required is high such as reinforced concrete structures, concrete grade is M25 and above, cement grouts, bridges, runways, and roads. It complies with the specifications set out in BIS 12269 standard (BIS 8112, 2013).

There are a variety of types of cement available. Portland pozzolana cement (PPC) is made by inter-grinding and mixing of OPC with fly ash, Portland slag cement (PSC) is made by mixing OPC with Ground Blast Furnace Slag, and Portland composite cement (PCC) is made by mixing OPC with Fly Ash, GBFS, etc. Portland limestone cement (PLC), Sulphate-resistant Portland Cement, Masonry Cement, Oil Well Cement, High-Alumina Cement, Super Sulphated Cement, Rapid Hardening Portland Cement, White Portland cement, etc., are additional types of cement.

The detailed process of manufacture of cement is explained in Chap. 4. The quality of cement is controlled as per the applicable standards. These standards are prescribed by the authorized agencies and/or standard institutions in the countries. In India, the same is done by the Bureau of Indian Standards (BIS).

Various standards of countries published for different grades of cement are listed in Table 5.1

The quality and composition of raw materials utilized in the cement manufacturing define its cement quality. Input control based on the quality of raw materials, therefore, is an important aspect of cement manufacture. Also, for effective quality control, a representative sample ensures reliable results and hence sample preparation is an important aspect of quality control.

Table 5.1 Standards for cement products from different countries

Country	Cement Product	Standard
India	33-Grade OPC	IS 269 (2013)
	43-Grade OPC	IS 8112 (2013)
	53-Grade OPC	IS 12269 (2013)
	Portland slag cement (PSC)	IS 455 (1989)
	Portland pozzolana cement (PPC)	IS 1489 (1991)
	Portland composite cement (PCC)	IS 16415 (2015)
UK & Europe	Different types of Cement	EN 197–1
USA	OPC	ASTM C150
	Blended cements	ASTM C595
Australia	Portland and Blended Cements	AS 3972

Table 5.2   Natural and Alternative materials supplying required elements in cement manufacture

Calcium	Iron	Silica	Alumina	Sulphate	Fuel
Natural Raw materials and Fuels					
• Limestone • Chalk • Clay • Marble • Marl • Sea shell	• Iron Ore • Clay • Shale	• Clay • Fly ash • Sand • Sandstone • Shale	• Bauxite • Cement rock • Clay • Fly ash • Fuller's earth • Shale	• Natural / Mineral Gypsum	• Coal • Petroleum Oil • Petroleum Gas • Petcoke • Lignite
Alternative Raw materials and Fuels					
• Alkali Waste • Slag • Lime sludge	• Mill Scale • Blast Furnace Flue dust • Ore washing • Red mud • Iron Sludge	• Slag • Ore washing • Rice-husk ash	• Copper slag • Ore washings Slag • Red mud	• FGD Gypsum • Chemical Gypsum	• Agro-waste • RDF • Plastic Waste • Hazardous Waste • Non-hazardous waste • Used Tyres and Rubber waste • Dried Sewage Sludge • Animal meal

5.2   Replacement of Conventional Materials with AFRs

The quality control of cement is achieved by performing tests on the raw materials, fuels, intermediate products, and final cement products. Depending upon the material, different kinds of tests are required to be performed.

There are five different elements that constitute the raw material in the cement manufacturing process, namely Calcium, Iron, Silica, Alumina, and Sulphate. Different natural and alternative raw materials that constitute these elements utilized in the cement manufacture are provided in Table 5.2.

5.3   Chemical Analysis Associated in Cement Manufacture

Table 5.3 provides details about the phases present in clinker and their major responsibility in cement chemistry.

Table 5.3 Clinker Phases in the cement

Chemical Name	Cement Notation	Mineral Name	Oxide Formula	Major Responsibility
Tri-calcium Silicate	C3S	Alite	$3CaO.SiO_2$	Early strength development
Di-calcium Silicate	C2S	Belite	$2CaO.SiO_2$	Later strength development
Tri-calcium Aluminate	C3A	Aluminate	$3CaO.Al_2O_3$	Setting time, initial strength increase, and heat release during setting
Tetra-calcium Alumino-ferrite	C4AF	Ferrite	$4CaO.Al_2O_3.$ Fe_2O_3	Clinker burning temperature reduction

In cement manufacturing, the analysis of raw materials and fuels is very important to determine the acceptability and process efficiency. The major focus of analysis of raw materials and fuels and their determinants is described in Table 5.4.

Three types of methods are used to undertake quality control of the raw materials, fuels, clinker, and Cement product.

1. Chemical analysis,
2. Physical analysis, and
3. Fuel analysis.

Table 5.4 Major focus of analysis of raw materials and fuels and determinants

Major focus of analysis of Raw materials	Determinants	Major focus of analysis of Fuels	Determinants
Main elements	CaO, Fe_2O_3, Al_2O_3, SiO_2, and MgO	Proximate analysis	Fixed Carbon, Moisture, Volatile Matter and Ash content, and GCV
Impurities	Water, hydrocarbons	Ultimate analysis	Carbon, Hydrogen, Nitrogen, Sulphur, and Oxygen
Circulation elements	Cl, SO_3, Na_2O, K_2O	Halogen content	Chlorine and Fluorine
Heavy metals	Hg, Cd, Tl, Pb, As, Cr	Heavy metals	Hg, Cd, Tl, Pb, As, and Cr
		Fuel value	Gross Calorific Value and Net Calorific Value
		Ash analysis	CaO, Fe_2O_3, Al_2O_3, SiO_2, and MgO

The chemical composition of raw materials and fuels is determined by two methods—Instrumental Techniques and Laboratory tests. Chemical analysis of the bulk material is also carried out using an online XRF analyser mounted on the crusher discharge belt or raw mill inlet belt. Instrumental Techniques involves X-Ray Fluorescence Spectrometer, X-Ray Diffraction, Optical Microscope, and Flame Photometer while the Laboratory test involves Gravimetric Method and Volumetric and Complexometric Methods. Short descriptions of each of these methods are given below.

a.   **X-Ray Fluorescence Spectrometer.**

XRF spectrometer provides a non-destructive analytical technique to determine the elemental composition of materials.

b.   **X-Ray Diffraction.**

X-ray Diffraction (XRD) is a rapid analytical technique used for phase identification of a crystalline material and can provide information on unit cell dimensions.

c.   **Optical Microscope.**

The optical microscopy technique shows a great many microstructural features of the clinker that are not visible using other microscopic techniques. These features include porosity changes, colour changes, constituent amounts (qualitative), individual crystals sizes and form, clustering size and form (alite, belite, and free lime), interstitial amounts and components, degree of crystallinity, minor constituents, and many other features.

d.   **Flame Photometer.**

A flame photometer is used to determine the concentration of certain metal ions like sodium, potassium, lithium, calcium, cesium, etc. In flame photometer spectra, the metal ions are used in the form of atoms.

e.   **Gravimetric Method.**

Gravimetric analysis is a method of quantitative chemical analysis in which the constituent sought is converted into a substance (of known composition) that can be separated from the sample and weighed.

f.   **Volumetric and Complexometric Methods.**

A complexometric titration is a form of volumetric analysis in which the formation of a coloured complex is used to indicate the endpoint of a titration. Complexometric titrations are particularly useful for the determination of a mixture of different metal ions in solution.

5.4 Cement Chemistry and Manufacturing Principles

Cement manufacture requires precise monitoring of the input materials based on their chemical constituents and also operational control. This control is well practised by the cement plants while utilizing conventional materials. The same precision is also required while utilizing AFRs. To achieve the desired quality of clinker and smooth process operation, it is important that the raw mix design with AFRs remains reasonably the same as the one designed with conventional materials. The chemistry of the ash from conventional fuel as well as AFR needs to be accounted for in the raw mix design. This aspect becomes critically more important while reaching higher TSRs.

5.4.1 Raw Mix Parameters

To achieve desired quality cement, desired quality clinker is required to be produced. This requires that the raw material mix having desired physico-chemical characteristics is utilized.

The design of this raw mix not only requires a complete analysis of raw materials but also requires the analysis of the ash content of the fuel as well. While undertaking co-processing, the analysis of both natural and alternative materials is important.

In raw mix design for the kilns undertaking co-processing, the composition of these different natural and alternative materials is arrived at based on several principles and fundamentals which are elaborated in detail below.

To illustrate the relevance of these fundamentals and principles, the case study data provided in Table 5.5 is utilized.

5.4.2 Raw Meal to Clinker Factor

$$F = \frac{100}{(100 - LOI)}.$$

LOI is the loss on the ignition of the raw mix.
Example: By utilizing the data provided in Table 5.1:
Raw meal to clinker factor = 100 / (100–35.2) = 1.54321.

5.4.3 Clinker Composition

Clinker composition is calculated from the raw mix, by calculating the constituent's composition on loss-free basis.

Table 5.5 Case study data on the composition of Raw mix

Parameter	Raw mix constituent
LOI	35.20
SiO2	14.00
Al2O3	3.70
Fe2O3	2.10
CaO	41.60
MgO	1.80
SO3	0.23
K2O	0.61
Na2O	0.22
TiO2	0.17
Mn2O3	0.30
P2O5	0.06
Cl	0.01
Total	100.0

This calculation is done by calculating the raw mix constituent with the raw meal to clinker ratio. The calculated values are provided in Table 5.6.

Table 5.6 Case study raw mix data converted into clinker data

Parameter	Clinker constituent
LOI	0.0
SiO2	21.6
Al2O3	5.7
Fe2O3	3.2
CaO	64.2
MgO	2.8
SO3	0.4
K2O	0.9
Na2O	0.3
TiO2	0.3
Mn2O3	0.5
P2O5	0.1
Cl	0.0
Total	100.0

5.4.4   Alumina Modulus AM

$$\text{Alumina Modulus} \ = \ \frac{(Al_2O_3)}{(Fe_2O_3)}.$$

This determines the potential relative proportions of aluminate and ferrite phases in the clinker. An increase in clinker AM means there will be proportionally more aluminate and less ferrite in the clinker.

The typical range of AM is $1.0 - 3.0$.

Example: By utilizing case study data:

Alumina Modulus $= 5.7/3.2 = 1.72$.

Higher Alumina Modulus indicates harder burning and entails higher fuel consumption. It also increases C3A, C3S, and C2S content and reduces C4AF content. Also, it reduces the liquid phase and kiln output.

If AM, < 1.23, Al_2O_3 acts as a flux, and if AM is > 1.23, Fe_2O_3 acts as a flux.

Very low Alumina Modulus causes sticky clinker and increases balling.

5.4.5   Silica Modulus SM

The Silica Modulus is defined as follows:

$$SM \ = \ \frac{SiO_2}{(Al_2O_3 + Fe_2O_3)}.$$

A high silica ratio means that more calcium silicates are present in the clinker and less aluminate and ferrite.

Higher Silica Modulus causes hard burning clinker and increases fuel consumption. This causes difficulty in coating formation and deteriorates the kiln lining. It results in slow setting and high strength cement.

Lower silica modulus increases the liquid phase. It improves the burnability of the clinker and the formation of coating in the kiln. SM typically ranges between 1.8 and 3.6.

Example: Utilizing case study data,

$$Silica\ Modulus \ = \ 21.6/\,(5.7 + 3.2) \ = \ 2.41.$$

5.4.6   Lime Saturation Factor (LSF)

The Lime Saturation Factor is a ratio of CaO to the other three main oxides. It is calculated based on the Alumina modulus number.

$$LSF = \frac{(CaO)}{(2.8 \text{ x } SiO_2 + 1.18 \text{ x } Al_2O_3 + 0.65 \text{ x } Fe_2O_3)}.$$

The LSF controls the ratio of alite to belite in the clinker. A clinker with a higher LSF will have a higher proportion of alite to belite.

Typical LSF values in modern clinkers are 85–100%.

Example: Utilizing case study data,

$$LSF = 64.2 / (2.8 * 221.6 + 1.1 * 5.7 + 0.7 * 3.2) = 93\%.$$

Values above 100 indicate that free lime is likely to be present in the clinker. This is because, in principle, at LSF = 100, all the free lime should have been combined with belite to form alite.

5.4.7 Hydraulic Modulus (HM)

The hydraulic modulus of good quality clinker is generally in the range of 2. Cement with HM less than 1.7 showed mostly insufficient strength and cement with HM more than 2.3 has poor stability of volume. With increasing HM, more heat is required to burn the clinker.

$$HM = \frac{CaO}{Al_2O_3 + Fe_2O_3 + SiO_2}.$$

Example: Utilizing the case study data,

$$HM = 64.2 / (5.7 + 3.2 + 21.6) = 2.1.$$

5.4.8 Bogue's Formula for Cement Constituents

The Bogue formulations are used to determine the compound compositions of the Portland Cement. These compounds have been researched by Rober Herman Bogue. Honouring his work, the primary compounds or constituents of the Portland cement responsible for the setting of cement paste are called Bogue's compounds. The ollowing are the four Bogue compounds. These are also the four main ingredients of cement. **These compound compositions may be calculated from the oxide analysis as follows (C, S, A, F, and $\underline{S}$ denote the % of CaO, SiO_2, Al_2O_3, Fe_2O_3, and SO_3, respectively):**

$$C3S = 4.071 \times CaO - (7.602 \times SiO_2 + 6.718 \times Al_2O_3 + 1.43 \times Fe_2O_3 + 2.852 \times SO_3).$$

$$C2S = 2.867 \times SiO2 - 0.7544 \times C3S.$$

$$C3A = 2.65 \times Al2O3 - 1.692 \times Fe_2O_3.$$

$$C4AF = 3.043 \times Fe_2O_3.$$

Typical value of

$$C3S = 45 - 55\%.$$

$$C2S = 20 - 30\%.$$

Example: Utilizing the case study data, we arrive at the following values.

$$C3S - 53.1\%.$$

$$C2S - 21.88\%.$$

$$C3A - 9.65\%.$$

$$C4AF - 9.86\%.$$

5.5 Operational Parameters

The salient features of some of the important operational parameters in the cement manufacture are discussed next which have relevance in the implementation of AFR co-processing successfully.

5.5.1 Degree of Calcination

$$C\,(\%) \;=\; \frac{(\text{fi}\;-\;d_i)}{(100\,/\,\text{fi})}$$

(Or)

$$C\,(\%) \;=\; \left(1-\text{LOI}_{sample}\right)\times(100-\text{LOI}_{feed})\,/\left(\left(100-\text{LOI}_{sample}\right)\times(\text{LOIf}_{eed})\right).$$

C: Apparent Percent calcination of the sample.
f_i: Ignition loss of the original feed.
d_i: Ignition loss of the sample.

5.5.2 Loss on Ignition

Loss on ignition (LOI) refers to the reduction in weight that takes place when the material is heated to a high temperature. This loss in weight is due to the release of $CO2$, Moisture, and burning of the combustible matter.

$$\text{LOI} \;=\; 0.44\,CaCO_3 + 0.524\,MgCO_3 + \;\ldots.+$$
$$\text{combined}\;H_2O \;+\; \text{Organic matter.}$$

5.5.3 % Liquid

$$\%\,\text{Liquid} \;=\; 1.13\,C3A + 1.35\,C4AF + MgO + \text{Alkalis.}$$

$$C3A : \%\;\text{of TriCalcium Aluminate.}$$

$$C4AF : \%\;\text{of Tetra}-\text{Calcium Alumino Ferrite.}$$

Example: By utilizing the case study data and the values calculated,

$$\%\,Liquid - 1.13*9.65 + 1.35*9.86 + 2.77 + 0.55 + 0.31 = 28.27\%.$$

5.5.4 Sulphur to Alkali Ratio

$$SO_3 / \text{Alkali(in absence of Chlorine)} = \frac{(SO_3/80)}{((K_2O/94) + (0.5Na_2O/62))}.$$

$$SO_3 / \text{Alkali(In presence of Chlorine)} = \frac{(SO_3/80)}{((K_2O/94) + (Na_2O/62) - (Cl/71))}.$$

The typical sulphur to Alkali ratio in absence of Chlorine is around 1.1 and the same in presence of Chlorine is 0.8. Preheater build-ups are the major cause of concern while operating at higher sulphur to alkali ratio.

Example: Assume SO3 content in the hot meal is 1.0% and K2O as 0.7%. Na2O is 1% and chlorine is 0.8%,

$$Sulphur\ to\ Alkali\ ratio = (1.0/80) / ((0.7/94) + (01/62) - (0.8/71)) = 1.02.$$

5.5.5 Free Lime

% Free Lime$_{1400}$ = 0.31 (LSF – 100) + 2.18 (SM – 1.8) + 0.73 Q + 0.33 C + 0.34 A.

LSF: Lime saturation factor.

SM: Silica modulus.

Q: +45 μ residue after acid wash (20% HCl) identified by microscopy as quartz.

C: +125 μ residue which is soluble in acid (i.e., coarse LS).

A: +45 μ residue after acid wash identified by microscopy as non-quartz acid insoluble.

Note: Q, C, and A expressed as % of total raw mix sample.

5.5.6 Excess Sulphur (Gm SO₃ / 100 Gm Clinker)

$$\text{Excess sulphur} = (1000 \times SO_3) - (850 \times K_2O) - (650 \times Na_2O).$$

Sulphur is required to be maintained between 250 gms / T clinker to 600 gms / T clinker. Coating problems are encountered in the preheater tower above these limits.

5.5.7 Blending Ratio

The blending ratio is the ratio of estimated standard deviations of feed and product.

Blending ratio = standard deviation of CaO in feed / standard deviation of CaO in the product.

= Sqrt (N/2)

N : Number of layers.

For calculating standard deviation,

Consider the feed values: $x, x_1, x_2, x_3 \ldots \ldots x_n$.

Mean for the feed values: $(x + x_1 + x_2 + x_3 \ldots x_n)/n = x_a$.

Standard deviation for the feed:

$= \text{sqrt}\{[(x-x_a)^2 + (x_1 - x_a)^2 + (x_2 - x_a)^2 + \ldots + (x_n - x_a)^2]/n\}.$

5.5.8 Ash Absorption

Ash Absorption = specific fuel consumption x % of ash in fuel.

$$\text{Specific fuel consumption} = \frac{\text{Kg coal}}{\text{Kg clinker}}$$

$$= \frac{\text{Specific heat consumption}}{\text{NCV of coal}}.$$

Note: Clinker is assumed to have zero LOI.

5.5.9 Kiln Feed to Clinker Factor

$$\text{Kiln feed to clinker factor} = \frac{\text{Kiln feed (kg)}}{\text{Clinker output (kg)}}.$$

Note: Error in kiln feeding system is considered as negligible.

(or)

$$\text{Kiln feed to clinker factor} = \frac{\text{Raw Meal to Clinker Factor} \times (100)}{\text{Top Stage Cyclone Efficiency}}.$$

5.5.10 Clinker to Cement Factor

Clinker to cement factor

$$= \frac{(\text{Clinker} + \text{Gypsum} + \text{additives} + (\text{Fly ash / slag}))(\text{Kg})}{\text{Clinker consumed (kg)}}.$$

5.5.11   Insoluble Residue

The material remaining after the cement is treated for designed time with hydrochloric acid of a specific concentration.
(Or)
It can be used to measure the amount of adulteration of cement with sand. Sand is insoluble and cement is soluble in dilute HCl. In PPC cement, this parameter provides the estimate of the percentage of the fly ash present in it.

5.5.12   Volumetric Loading of Kiln

$$\text{Volumetric Loading } (\text{TPD/m}^3) = \frac{CP}{\Pi \times (D^2/4) \times L}.$$

CP = Clinker Production in TPD.
D = Effective Diameter of the kiln (n).
L = Length of kiln (m).
Typical values of Specific volumetric loading for preheater kilns range from $1.6 - 2.2$ tpd/m^3.
And for the pre-calciner kilns, it ranged from 4.5 to 7.0 tpd/m^3.

5.5.13   Thermal Loading of Kiln

$$\text{Thermal Loading } (\text{MKcal/Hr./m}^2) = \frac{CP \times HC \times \% \text{ Firing in kiln} \times 10^3}{\Pi \times (D^2/4) \times 24 \times 10^6}.$$

CP = Clinker Production (tpd).
HC = Heat Consumption (Kcal / Kg).
D = Effective Diameter of the Kiln (m).
The specific thermal loading for pre-calciner kilns ranges from 4.0 to 5.0 M kcal/hr./m^2.

5.5.14   Feed Moisture Evaporation Rate

Feed moisture evaporation rate
Moisture (kg/hr) = $F_q \times 1000 \times (M_f - M_p) / (100 - M_f)$.
F_q: Fresh feed quantity (tph).
M_f: Total fresh feed surface moisture (%).
M_p: Total product surface moistures (%).

5.5.15 False Air Estimation O_2 Method

$$\text{In terms of outlet}:\ X\ =\ \frac{(O_2(\text{outlet}) - O_2(\text{inlet}))\ \text{x}\ 100\ (\%)}{(21 - O_2(\text{inlet})}.$$

$$\text{In terms of inlet}:\ X =\ \frac{(O_2(\text{outlet}) - O_2(\text{inlet}))\ \text{x}\ 100\ (\%)}{(21 - O_2(\text{outlet}))}.$$

5.5.16 % Excess Air

% Excess air calculation is carried out based on two criteria. With zero CO, i.e., complete combustion and with a certain level of CO which is incomplete combustion.

Case 1: For Complete combustion

$$\%\ \text{Excess air}\ =\ O_2 / (21 - O_2).$$

Case 2: For Incomplete combustion (with CO)

$$\%\ \text{Excess air}\ =\ 189\ \text{x}\ (2O_2 - CO) / (N_2 - 1.89\ \text{x}\ (2O_2 - CO)).$$

5.6 Thermal Parameters

The following thermal parameters utilized in cement, manufacture are also important while implementing AFR co-processing.

A. *Conversion of GCV to NCV*

$$\text{NCV}\ = \text{GCV}\ -5150 * \text{H (kcal/kg) where H is the \%}$$
$$\text{Hydrogen present in fuel and moisture).}$$

B. *Ultimate Analysis*

Ultimate analysis of fuel is a measure of C, H, N, S, O, and Ash and the sum of all these is 100% (by weight) where C is % carbon, H is % Hydrogen, N is % nitrogen, S is % sulphur, and O is % oxygen.

This parameter is useful to calculate the volume of combustion gases and theoretical combustion air required.

C. *Proximate Analysis*

The proximate analysis of fuel is the quantitative determination of moisture, carbon, volatile matter, and ash. It is the sum of % volatile, % fixed carbon, % ash, and % moisture and is equal to 100%.

D. *% Coal ash absorbed in clinker*

$$X_1 = (CaO_{\text{Clinker}} - CaO_{\text{raw mix}})/(CaO_{ash} - CaO_{raw\,mix}).$$
$$X_2 = (Fe_2O_{3\,\text{Clinker}} - Fe_2O_{3\,\text{raw mix}})/(Fe_2O_{3ash} - Fe_2O_{3\,raw\,mix}).$$
$$.X_3 = (SiO_{2\,\text{Clinker}} - SiO_{2\,\text{raw mix}})/(SiO_{2ash} - SiO_{2\,raw\,mix}).$$
$$X_4 = (Al_2O_{3\,\text{Clinker}} - Al_2O_{3\,\text{raw mix}})/(Al_2O_{3ash} - Al_2O_{3\,raw\,mix}).$$
$$\%\ Coal\ ash\ absorbed\,in\,Clin\ker = (X_1 + X_2 + X_3 + X_4)\,/\,4.$$

5.7 Burner Parameters

Burner is important equipment in the successful implementation of AFR co-processing. The relevance of these thermal parameters in AFR co-processing operation needs to be understood and tackled appropriately.

A. *Theoretical air required to burn fuel*

$$\text{Air (kg air/kg of fuel)} = (8/3) \times C + 8 \times (H_2 - (O_2/8)) + S) \times (100/23).$$

C: Mass of carbon per kg of fuel.
O_2: Mass of Oxygen per kg of fuel.
S: Mass of Sulphur per kg of fuel.
H_2: Mass of hydrogen per kg of fuel.
Example: The Carbon content in the fuel is 53.28%, Oxygen is 15.8%, Sulphur is 0.13%, and Hydrogen is 5.54%.

$$Theoretical\ Air\ required = (8/3) * (53.28/100)$$
$$+ (8 * ((5.54/100) - ((15.8/100)/8)) + (0.13/100)) * 100/23$$
$$= 2.71\ Kg\,air\,/\,Kg\,fuel.$$

B. *Primary air momentum*

Primary air momentum is calculated (% m/sec) as follows:
% m/s = L_p % x C,

where
L_p : The primary air represented as % of the stoichiometric air required.
C : Primary air velocity at the burner nozzle.

Example: Primary airflow rate is 2.14 kg/sec; the stoichiometric air required is 22.17 kg/sec. The nozzle velocity at the burner is 165.6 M/sec.
The primary air % = 2.14/22.17 = 9.6%.
The Primary air momentum = 9.6*165.6 = 1598 % M/s.

C.   *Burner momentum*

The burner momentum (N/MW) is calculated in the following manner:

$$\text{Burner Momentum} = \frac{\text{Nozzle velocity (M/s) x Primary airflow rate (kg/s)}}{\text{Total heat input in the kiln (MJ/s)}}.$$

Example: The nozzle velocity is 165.6 m/s, the primary airflow rate is 2.14 kg/s, and the total heat input in the kiln is 65.7 MJ/s.

$$Burner\ Momentum = 165.6 * 2.14/65.7 = 5.39\ N/MW.$$

D.   *Nozzle velocity*

Nozzle velocity is calculated using the following equation.
Nozzle Velocity:

$$C_{pr} = \sqrt{\frac{2k}{k-1} \times R\,(t_{pr} + 273.15) \times \left[1 - \left(\frac{P_{amb}}{P_{amb} + P_N}\right)^{\frac{k-1}{k}}\right]}\ [m/s],$$

where
P_{amb} (mbar) pressure at Nozzle.
t_{pr}Primary Air Temperature.
K~ 1.4 Isentropic exponent for air.
R~Gas constant (286.89 (J/kgK)).
C_{pr}Nozzle Velocity (m/s).

Example: *Primary Air temperature = 60 Degree C.*
Primary pressure at nozzle = 242 mbar.
Ambient pressure = 972 mbar.
K = 1.40.
R = 286.89 J/kg k.
Putting all these values in the above formula for calculating nozzle velocity,

$$Nozzle\ Velocity C_{pr} = 165.60\ m/s.$$

E.   *Estimated*Burner Nozzle velocity (v)

$$v \quad \sim \quad \sqrt[4]{P_s}\ m/\sec\ (P_s\ in\ mmWC),$$

$$v \quad \sim \quad \sqrt{\frac{200 \times P_s}{\rho}} \; m/\sec \; (P_s \; in \; mbar),$$

where P measured at the axial air pressure point.

p measured as Density.

$$V = 4 \, X \, S_{QRT}(primary \, air \, static \, pressure).$$
$$Primary \, air \, pressure = 1600 \, mmwg \quad Normal \, Coal.$$
$$Primary \, air \, pressure = 2420 \, mmwg \quad Petcoke.$$
$$Nozzle \, Velocity = 4 \, X \, S_{QRT}(1600)$$
$$= 160 \, m/s \quad for \, normal \, coal$$
$$= 4 \, X \, S_{QRT}(2420)$$
$$= 196.76 \, m/s \quad for \, petcoke.$$

F. *Position of Burner in Cement Kiln*

The position of the burner in the kiln and the flame profile inside the kiln shell play an important role in achieving the optimum level of heat transfer to the material for obtaining the desired cement-related reactions without impacting the refractory lining. This concept is depicted in Fig. 5.1.

The optimum position, therefore, depends on many factors. (Hegde, 2020). In the past, it was a common practice to point the burner a little bit down compared to the kiln axis, in direction of the charge. This was primarily done to compensate for the tendency of the flame to go upwards due to convection and entrainment by the secondary air.

Today, with modern high-momentum burners, this is no longer recommended. The jet momentum being stronger, if you point the burner towards the charge, the risk is that the flame will touch the charge. The local reducing conditions would increase sulphur circulation and increase the risk of coating and blockages in the

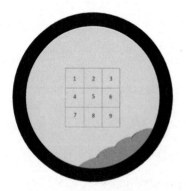

- Position 1,2,3,4 & 7 are close to the refractory lining and may impact its performance.
- Position 6, 8 & 9 are close to the process material and may cause over heating the same.
- Position 5 is the best because it is placed away from process material and also from the refractory and provides opportunity for optimum performance

Fig. 5.1 Position of the burner in the cement kiln

preheater. The consensus is that high-momentum burners should be placed parallel to the kiln axis. Basic positions would be on the kiln axis, but the burner can also be shifted sideways (still parallel to the kiln axis). Some recommend shifting the burner horizontally away from the charge when using coarse waste fuel to limit the risk that coarse particles would fall into the charge. Similarly, if you use only fine, easy-to-burn fuel, the burner can be shifted towards the charge to improve heat exchange. As radiation is the primary heat exchange mechanism, the effect is, however, limited.

Regarding insertion depth of burner inside the kiln, according to theory, the further inside is the better. This is to get away from the perturbation of the change of direction of the secondary air and to improve the precooling zone to avoid snowmen in the cooler. There is, however, a limit due to the length and related weight of the burner and the risk of damage by big pieces of coating falling on the burner. *The usual insertion depth would be 50 cm–1 m inside the kiln.* But many kilns operate with the burner just at the limit of the kiln (0 cm). Having the burner outside the kiln is generally not recommended.

5.8 Conclusions

Utilization of AFRs through co-processing requires alignment of the chemistry of AFRs with that of the input materials and design of the raw mix to achieve a desired clinker composition. The objective is to ensure that the final chemistry of the resultant mix of fossil materials and alternative materials remains the same as is required to produce a desired quality of clinker. Further, it is also desired that the undesired components entering through the alternative materials are curtailed to the levels that permit trouble-free and environmentally sound process operation. This requires an understanding of the various aspects of the cement chemistry as well as the operational and quality control in the cement manufacturing process. This chapter dealt in detail with the understanding of these aspects with case study examples using typical data.

References

CII-Godrej GBC, Cement Formulae Handbook published by CII - Sohrabji Godrej Green Business Centre. (2010).
BIS, ORDINARY PORTLAND CEMENT, 33 GRADE—SPECIFICATION, IS 269, Bureau of Indian Standards. (2013, Mar).
BIS, ORDINARY PORTLAND CEMENT, 53 GRADE—SPECIFICATION, IS 12269, Bureau of Indian Standards. (2013, Mar).
BIS, ORDINARY PORTLAND CEMENT, 43 GRADE—SPECIFICATION, IS 8112, Bureau of Indian Standards. (2013, Mar).
Dr. Hegde, S. B. (2020, February 7). Position of Burner in Cement Kiln, Linkedin post.

Part IV
Co-processing: Guidelines, Sustainability and Legislation

Chapter 6
Guidelines on Pre-processing and Co-processing of AFRs—International Best Practices

6.1 Introduction

Co-processing of AFRs is being practised globally for about four decades. Over this long period, there have been many learnings and experiences in undertaking co-processing initiatives. Many best practices have come in place in the process. These learnings and best practices have helped build appropriate guidelines. These guidelines, best practices, and learnings are discussed in detail in this chapter. The most important principle of these guidelines is to ensure that the unsuitable and unacceptable wastes are prevented from being co-processed and that emissions from co-processing activity do not get influenced by the co-processing activity.

Common concerns of AFR co-processing in cement kilns are the following:

- Spills, accidents, and exposure during handling.
- Emissions during handling/pre-processing and co-processing.
- Contamination of the product.

This guideline elaborates the international best practices employed to achieve environmentally sound management of the AFRs / Wastes (GTZ-Holcim, 2006; IPPC, 2013; UNEP, 2007); World Business Council for Sustainable Development (WBCSD), 2006). When the same are required to be practised elsewhere, they need to be adopted to the local considerations. The local raw material and fuel chemistry; availability of AFRs and waste materials; infrastructure available at the cement production plant; availability of equipment for controlling, handling, and feeding the waste materials; and site-specific health, safety, and environmental issues are typical local considerations that need to be addressed. Some of these considerations would be site-specific and will be varying from plant to plant. Hence, these recommendations need to be perceived as general ones. The local permit, therefore, must specify the final details of the pre-processing and co-processing practices that would be allowed. The following requirements need to be put in place by the cement plant or the agency handling the waste / AFR materials (Dahai Yan et al., 2014; Karstensen et al., 2014; Karstensen, 2014, 2011; Karstensen et al., 2010; Karstensen, 2008;

© The Author(s), under exclusive license to Springer Nature Singapore Pte Ltd. 2022
S. K. Ghosh et al., *Sustainable Management of Wastes Through Co-processing*,
https://doi.org/10.1007/978-981-16-6073-3_6

Karstensen et al., 2006; Karstensen, 2006). These requirements prevent or reduce the risks associated with them—especially the hazardous ones. The following requirements have been categorized for easy understanding which need to be complied with or adhered to. Table 6.1 demonstrates the requirements.

Table 6.1 Requirements by the cement plant or waste / AFR materials handling agency

Categories of requirements	Requirements
Compliances to international, national, and local Regulation Legislation and conventions	Approved EIA and applicable national and local permits Regulatory compliance at national and local levels Performance of the BAT/BEP and compliance to the relevant conventions such as Basel and Stockholm
Facility approval and Reliable Resource support	Facilities for pre-processing and co-processing that are approved Reliable power and water supply Appropriate laboratory and equipment facilities
Pollution including emission control	Appropriate pollution control devices and continuous emission monitoring systems at stacks for ensuring compliance to relevant regulations Preventing dioxin and furan formation by conditioning / cooling the exit gas to below 200 °C quickly
Calibration	Calibration of measuring and monitoring equipment and instruments regularly through baseline monitoring processes
Trial Management	Demonstration of destruction performance through co-processing trials
Training and Competence of personnel on technology, HSE, and emergency response	Personnel with required skills to manage health, safety, and environmental issues of the hazardous wastes Regular training of the personnel to handle safety equipment, procedures, and emergencies
Responsibility matrix	Well-defined organizational structure with clearly defined responsibilities
Defined acceptance, quality control, and authorization procedure	Safe receipt, storage, pre-processing, and co-processing of hazardous wastes Authorization for collection, transportation, and handling of hazardous wastes Waste acceptance and co-processing control through appropriate laboratory and equipment facilities Adequately defined control procedures for product quality

(continued)

Table 6.1 (continued)

Categories of requirements	Requirements
Records keeping and reporting	Maintenance of the records of hazardous wastes and emissions Regular disclosure of performance status through open reports Reporting system for employees to efficiently deal with errors
Certification of management systems	Implementation of ISO 9001, ISO 14001, ISO 50001, ISO 45000, EMAS, or similar systems to ensure quality, environmental and energy management, occupational health and safety, and continuous improvement
Complaints handling system and Auditing	Regular audits through independent agencies and emission monitoring and reporting through third-party agencies Engagement with local community and authorities to address comments and complaints

6.2 Important Operational Aspects

While starting to undertake co-processing in cement plants, certain measures and considerations need to be addressed. The following aspects are very important to look into and for taking appropriate actions.

(a) Compliance with regulations:

An appropriate and effective regulatory framework should be established to guarantee high-level protection of impacts on environment. It is suggested that a register of applicable legislation with acceptance limits is maintained and all the relevant data are entered into it in each shift. Threshold limits must be set for each of the critical parameters. A system of indicative measures needs to be installed which will help in taking appropriate corrective and preventive actions so that the parameters so that the limiting conditions are never reached.

(b) Location, health, and safety aspects:

There will always be some concerns associated with community related to emissions, logistics, transport, infrastructure, etc. The feasibility of the location for undertaking pre-processing and co-processing activities needs to be carefully evaluated prior to deciding any location. This will reduce risks associated with proximity to the community. There will also be a need to put in place corrective measures to avoid the release of vapours, odours, etc. The pre-processing facility and cement plant operator must, therefore, design and put in place appropriate operational and management (O & M) procedures to deal with emergency situations so that the safety of neighbours, workers, and installations is addressed appropriately; these needs to be systematically reviewed and revised.

(c)   Training:

A training programme for employees to address operation, safety, health, environment, and quality issues needs to be implemented and the associated training records need to be filed.

(d)   Involvement and communication:

Before starting to use any wastes, adequate documentation, and information about is safe handling and management are mandatory must be provided to all employees. Relevant authorities must be given access to relevant information about the waste material during the permitting process. Further, the operator of the pre-processing facility and cement plant must do the following:

- By having transparent, continuous, and consistent communication with the authorities and other involved stakeholders, necessary trust will be created. Stakeholders must have access to relevant information to allow them to clearly understand the purpose of the co-processing activity.
- Ensure that authorities can evaluate the entire process by providing them with all the necessary information.
- By working with the local community and authorities, the operator must establish a stakeholder engagement plan, which provides procedures for responding to community's interests, comments, or complaints.

(e)   Reporting performance:

Transparency and accountability are important criteria for building trust among stakeholders. Updated performance reports will ensure a fair and balanced judgment of the co-processor or the activities at the site and its performance.

(f)   Environmentally sound management:

Hazardous wastes defined within the Basel and Stockholm Convention need to be brought under environmentally sound management (ESM) policy concept.

(g)   Environmental management system:

For ensuring continuous improvement in its performance, the pre-processing facility should have a environmental management system (EMS) in place. The same is expected with the cement plant as well. ISO 14001 and the European standard—EMAS—are the two most frequently used international standard guidelines for EMS design.

6.3 Waste Evaluations

The cement plant operator must specify the waste acceptance criteria in advance to the pre-processing facility or the waste owner prior to any deliverables. These outer boundaries will define the acceptance at a particular kiln and then facilitate the definition of the requirements to comply with the delivery specification. Waste fed to a cement kiln should preferably be homogenous and have stable heat and moisture content and have a pre-specified size distribution with stable chemical and physical composition. In actual practice, the co-processor usually receives wastes having varying physico-chemical characteristics. To fulfil the requirements mentioned above, wastes may need to be pre-treated or pre-processed. Points needed for the waste evaluation process should include incoming waste evaluation, assessment of possible impacts, banned wastes, risky wastes, and Checklist for acceptance control which are explained below and those should be adopted in the facility.

(a)   Incoming Waste evaluation:

Wastes need to be received from parties that can provide traceability prior to the reception and unsuitable wastes need to be quarantined and refused. The pre-processing facility and cement plant operator must develop a joint evaluation and acceptance procedure for the waste.

(b)   Assessment of possible impacts:

After receiving the waste-related information, the co-processor should adopt the following procedures.

1. Evaluate any negative impact of handling and co-processing the wastes on the safety and health of the employees, contractors, and the community.
2. Identify suitable PPEs that facilitate safe handling of the waste at the site and provide them to the employees.
3. Ensure that non-compatible wastes are not mixed.
4. Assess the physico-chemical characteristics of the wastes and evaluate the impact the waste may have on the process operation due to the presence of alkalis, chlorine, sulphur, and fluorine content in the wastes. These may build up in the kiln system, leading to process concerns such as unstable operation and clogging. The calorific value defines the contribution waste makes to the energy requirement while the moisture content may reduce the efficiency and the productivity of the kiln system. The chemical composition of the cement will be influenced by the ash content in the waste and its chemical composition needs to be suitable or adjusted in the raw mix.
5. Understand the potential impact of the waste on the quality of the final product and the process stability.
6. Understand the impact that waste may have on emissions.

7. Define the quality control parameters of the waste that the supplier will have to adhere to. Each delivery must comply with the same prior to acceptance at the site.

(c) Banned wastes:

The following are the banned waste materials for co-processing. Some of them are banned materials both for pre-processing and co-processing. Local circumstances, statutory conditions, and company policy may cause some companies to exclude additional materials.

Materials Banned for pre-processing and co-processing

1. Infectious and biological active medical waste.
2. Explosives.
3. Asbestos.
4. Radioactive waste.
5. Unknown/unidentified wastes.

Materials Banned for co-processing after pre-processing

1. Entire Batteries.
2. Electronic waste.
3. Mineral acids and corrosives.
4. Unsorted municipal waste.

(d) Risky wastes:

The following are the waste streams that pose risk in the cement manufacturing process and may be restricted by some of the cement plants. For some waste streams, specific packaging may be required to prevent any potential reaction of wastes during transport.

1. Wastes containing mercury and thallium.
2. Waste containing peroxides.
3. Wastes forming acid gases during combustion.
4. Wastes containing phosphides.
5. Wastes containing isocyanates.
6. Wastes containing Cyanides.
7. Wastes with alkaline metals or other reactive metals.

(e) Checklist for acceptance control:

The operator must make sure that the following criteria are agreed prior to signing any commercial contract. Figure 6.1 illustrates some typical waste acceptance criteria.

1. Adequate information on the risks associated with the material along with its composition and an agreement to deliver the same through a declaration.

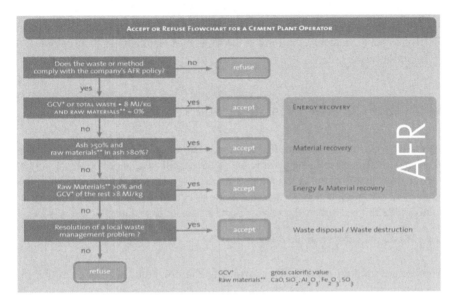

Fig. 6.1 Waste acceptance criteria

2. No substances, compounds, or preparations which are on the "negative list" or not agreed for acceptance should be sent.
3. Each waste must be tested for its compliance with the agreed waste characteristics. After comparison of the results of testing of the waste materials brought for delivery with the data contained in the declaration, the waste is accepted at the cement plant or sent to an appropriate pre-processing facility or rejected in the case of significant deviations.
4. Blending of incompatible materials is prohibited and a compatibility test is performed by the process owner, if needed.
5. Before acceptance of commercial contracts, the plant shall carry out sampling of the waste at the site of the provider and tested for its physico-chemical characteristics. This analysis can be carried out by own- or third-party certified laboratories.
6. Before completion of the acceptance process, cement plants should not permit the transportation of the wastes to the plant site.
7. To ensure that PPEs and safety gears are utilized appropriately, cement plants shall communicate the health risks and the safety concerns to the personnel associated with the downstream operations, including transportation, pre-processing, and co-processing.
8. Personnel handling the waste needs to be provided SOP of materials handling based on the material properties and available safety information. MSDS (Material Safety Data Sheet) or any such document must be made available at the workplace and knowledge and facility of implementing MSDS must be provided.

9. To facilitate enforcement of the waste acceptance criteria, the cement plant shall provide adequate training in the subject of chemistry to their commercial employees.

6.4   Waste Collection, Handling, and Transport

Standard procedure for collection, handling, and transport of wastes to the cement plant and the pre-processing facility must be established, monitored, and evaluated in full compliance to the national regulatory requirements. Qualified, authorized by respective local and/or national regulatory authority, and licensed transport companies shall be used as the service provider for transportation of the wastes.

(a)   Waste collection and handling:

Appropriate training needs to be provided to the operating manpower at the pre-processing and co-processing facilities. The hazardous and other wastes must be handled separately. This will prevent intermixing and contamination.

(b)   Waste transport:

Packaging of wastes that are hazardous in nature is desired before undertaking their transportation. This is true even for wastes contaminated with hazardous materials. The material stored in containers must meet transport requirements so that when they are transported, they meet the applicable requirements.

Figure 6.2 illustrates the transport vehicle carrying well-stacked drums containing waste

6.5   Waste Reception and Handling

Waste material must be packed, labelled, and loaded properly during transportation. This will ensure that waste material will safely reach the pre-processing and co-processing facility. Detailed instructions on the types of waste material packed in the drums is mentioned properly on the drums. Compliance to the specifications confirms that the waste is positively verified. Therefore, all wastes should be evaluated for this compliance. Until then it should be treated as unknown and hazardous. The procedure to be followed for accepting the vehicle and he waste contained in the same shall be as follows.

(a)   Vehicle must be stopped on its arrival and necessary identification has to be made.
(b)   Weight of vehicles should be taken and recorded at the site incoming and exit gate.

Fig. 6.2 Transport vehicle carrying well-stacked drums containing waste

(c)   Documents relating to hazardous waste must be checked. Verification must be done for compliance with site acceptance specifications and applicable regulations.

(d)   Waste certificates, transport certificates, and other such certifications must be checked at the site.

(e)   Waste material must be sampled and tested to check compliance with the accepted quality criteria.

(f)   The truck driver must be properly made aware about safety and emergency requirements pertaining to the waste being transported. He should also be made aware of the unloading instructions. This procedure must be ensured before any of the materials are loaded into the truck.

(g)   Non-complying vehicles should not be permitted to enter the site.

6.5.1 Management of Non-Compliant Deliveries

The contractual agreement for the supply of wastes should accompany the SOP to be adopted on the non-compliant deliveries by the waste producer. In case of conflict, there should be agreement on criteria for rejection of the received vehicle along with the material contained in it. Where relevant, the rejection must be communicated to the relevant authority. Statistical analysis of the performance and reliability of each waste producer must be carried out based on the waste acceptance criteria and the records must be kept for a definite record retention period; their contracts also need to be reviewed periodically. These performance data should be considered as inputs for rating the waste producers/suppliers for making the AVL (Approved Vendor List) for further orders.

6.5.2 Analysing Incoming Wastes

Delivered wastes must undergo specific admission controls and must be confirmed to comply with the agreed specifications. The techniques for checking, inspection, and detection of materials are described below.

(1)   Techniques for checking wastes:

Different techniques varying from simple to complete chemical analysis are employed for checking the waste streams. The procedures adopted will depend upon the following.

(a)   Experiences of dealing with the wastes.
(b)   Nature and composition of waste.
(c)   Existence or absence of a quality specification of the waste.
(d)   Heterogeneity of the waste.
(e)   If the waste is of a known or unknown origin.
(f)   Known difficulties with wastes.
(g)   Specific sensitivities of the installation concerned.

(2)   Inspection:

The following inspection scheme is applied for the waste.

(a)   Assessment of flash point of wastes in the bunker and its control.
(b)   Comparison of data in the declaration list with delivered waste.
(c)   Blending tests on liquid wastes prior to storage.
(d)   Sampling of all bulk tankers.
(e)   Screening for elemental composition.
(f)   Assessment of combustion parameters.
(g)   Checking of drums randomly.
(h)   Checking of packaged loads.

(3)   Detectors for radioactive materials:

Operational and safety problems will be faced due to higher levels of radioactive sources or substances in waste. The same needs to be checked with appropriate detectors and controlled.

6.5.3 Reception and Handling

For the unloading, storage and handling of liquid and solid wastes, there should be written procedures and instructions available at the site. These must include following.

(a) Clearly designated routes within the site for vehicles carrying specified hazardous wastes.
(b) Training of relevant employees in the company's operating procedures and implementing regular auditing of these procedures for compliance.
(c) There should be appropriate signs at storage, stockpiling, and tank locations indicating the nature of hazardous wastes stored there.
(d) Control emissions to air, water, and soil from the storage facilities.

6.5.4 Labelling

With every delivery of the waste, its suitable description needs to be put up. Appropriate assessment of this description of the waste and assessment of the waste itself form a basic part of waste quality control. The following may be considered as the most important parameters for labelling.

(a) Origin of the waste.
(b) Name and address of the deliverer.
(c) Volume.
(d) Concentration of chlorides, sulphur, heavy metals, and fluorides.
(e) Calorific value.
(f) Water and ash content.

Figure 6.3 demonstrates the labelling of UN Dangerous goods.

6.6 Waste Pre-Treatment and Pre-Processing

To achieve continuous and smooth kiln operation, desired product quality, and no impact on the site's normal environmental performance, pre-processing of wastes to achieve more stable combustion conditions and homogeneous characteristics may be necessary.

Pre-treatment and pre-processing include shredding, drying, mixing, grinding, etc. These operations would be employed depending on the type of waste. Pre-processing may be located inside or outside the cement plant and is usually designed as per the waste characteristics. If the waste is prepared into alternative fuels outside the cement plant, these fuels only need to be stored at the cement plant. These are then proportioned suitably and fed to the cement kiln.

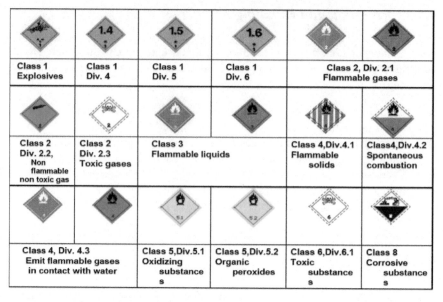

Class 1 Explosives	Class 1 Div. 4	Class 1 Div. 5	Class 1 Div. 6	Class 2, Div. 2.1 Flammable gases	
Class 2 Div. 2.2, Non flammable non toxic gas	Class 2 Div. 2.3 Toxic gases	Class 3 Flammable liquids		Class 4,Div.4.1 Flammable solids	Class4,Div.4.2 Spontaneous combustion
Class 4, Div. 4.3 Emit flammable gases in contact with water		Class 5,Div.5.1 Oxidizing substances	Class 5,Div.5.2 Organic peroxides	Class 6,Div.6.1 Toxic substances	Class 8 Corrosive substances

Fig. 6.3 UN Dangerous goods labelling

6.6.1   Types of AFRs

AFRs are of five types.

1. *Gaseous* such as refinery waste gas, landfill gas, pyrolysis gas, and coke oven gases.
2. *Liquids* such as hydraulic oils, low-chlorine spent solvents, insulating oils, distillation residues, and lubricating as well as vegetable oils and fats.
3. *Pulverize,* such as planer shavings, granulated plastic, ground waste wood, animal flours, sawdust, fine crushed tyres, agricultural residues, dried sewage sludge, and residues from food production.
4. *Coarse* such as crushed tyres, waste wood, re-agglomerated organic matter, and rubber/plastic waste.
5. *Lump* such as material in bags and drums, whole tyres, and plastic bales

The feeding and combustion behaviour of the wastes will be improved by mixing and homogenization. Mixing of wastes should only be carried out according to a known and documented recipe to avoid risks.

6.6.2 Pre-Processing of AFRs

There are several techniques that are utilized for waste pre-processing. These include the following:

(a) Mixing of liquid wastes to comply with the desired level of composition, heat content, and/or viscosity.
(b) Packaged wastes and bulky combustible wastes are processed through shredding, crushing, shearing, etc.
(c) Mixing of wastes in a bunker.
(d) Processing of source segregated combustible waste and/or other non-hazardous waste to produce refuse-derived fuel (RDF).

A bunker or a pit is utilized to mix the solid heterogeneous wastes prior to loading into the truck or sending to the feed systems.

6.6.3 Segregation of Waste Types for Safe Processing

The techniques utilized for segregating different wastes depend upon the waste type. To separate chemically incompatible materials, extensive knowledge is required; avoid mixing of non-hazardous and hazardous ones. Figure 6.4 depicts the guidance on the compatibility of chemicals.

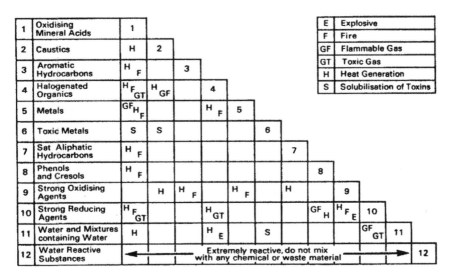

Fig. 6.4 Compatibility and reactions

6.6.4   General Design Considerations

All processing facilities need to ensure access to emergency escape routes, maintainability of the plant and equipment, and day-to-day operations.

A.   Design for reception and storage of hazardous wastes

The design of transfer and storage areas must take into account the accidental spills of the waste materials into rainwater or firewater and managing them. To avoid risks from fugitive vapour or spillage or emissions, safe material transfer systems to the storage area need to be put in place. Further, to minimize the impact of unloading activities, suitable vapour filtration and capture equipment should be implemented at the reception point and surrounding areas. This requires appropriate considerations for containment, isolation, and treatment. These are elaborated below.

(a)   Ensure that the spills will not penetrate the ground and contaminate soil and groundwater. For this, recommendations are to have properly designed drainage and sealed concrete surfaces.
(b)   All spills, leaks, and contaminated rainwater are properly collected and sent for disposal.
(c)   The runoff water from the storage area should get mixed with the sewers or stormwater drains. Such runoff should be designed for diverting them into a storage tank from where it can be sent for high-temperature destruction in the kiln.
(d)   Leak-free design should be implemented.
(e)   Piping leaks must be recovered without environmental contamination. For these, suitable methods to contain and recover them are designed.
(f)   Adequate alarms should be installed to identify abnormal conditions.

Periodic monitoring of VOC emissions should be carried out. To signal accidental waste fuel leaks, VOC detection should be placed at key process locations.

Volatile organic emissions from waste storage and pre-processing facilities should be exhausted into the hot zones of the cement kiln to achieve destruction. When loading the tank trucks, the vents of the storage tanks and the tank trucks should be connected through a closed vapour line to return the displaced VOCs from the storage tanks to the tank truck.

An activated carbon adsorption filter system could be provided as a backup to control VOCs from the storage tank. All the storage tanks should be equipped with explosion-proof safety valves.

B.   Housekeeping

General tidiness and cleanliness standards need to be exercised in the facility to achieve a good working environment. This also allows the identification of potential operational problems in advance.

6.6.5   Waste Storage

It is desired to evaluate maximum permissible waste storage possible based on the installed fire protection systems in the storage area and then limit waste volumes in storage to a level lesser than the same. The fire protection system should also include temperature and smoke detectors.

To consider the unknown nature and composition of wastes, assure that storage facilities fit their purpose. Storage time limits need to take local regulations, health, and safety risks into consideration.

Try to ensure that hazardous wastes are stored in the same containers (drums) that are used for transport, thus, avoiding the need for additional handling and transfer.

A.   Liquid and solid wastes

Hazardous wastes should be stored in an isolated area having an appropriate security system and protected from intruders. Incompatible wastes must be kept separate. The waste liquid storage sump area should be enclosed, and all gases from the sump area and storage tank should be vented to the VOC destruction system. Appropriate dust control systems should be implemented on the solid materials handling systems.

Liquid hazardous waste and sludges are usually stored in a series of tanks. Some tanks have storage under a Nitrogen atmosphere depending upon the flammable nature of the stored liquid. Liquid waste may be pumped to the kiln for co-processing using a suitably designed piping system. Sludges can be fed by using special "viscous-matter" pumps. Relevant safety and design codes must be utilized for the design of storage systems for the storage of liquid wastes based on operating pressures and temperatures. They also must have the adequate arrangement for secondary containment. Figure 6.5 depicts the good practices in stacking the drum packaged materials.

B.   Storage time

Storage of hazardous waste should be in accordance with the permit and regulation and for as brief a period as possible. Recommended storage times are as follows:

(a)   10 days for mixed wastes and hazardous wastes.
(b)   21 days for impregnated substrates.
(c)   For non-hazardous AFR, storage time is limited by the designed storage capacity and installed fire systems.

C.   Storage of solid waste

Solid that do not have smell can be stored temporarily in bunkers. The air in the bunker may be sent to the kiln using a duct. In addition to constant monitoring by personnel, heat-detecting cameras need to be used in locations where fires are anticipated.

Fig. 6.5 Storage of drummed waste at a pre-processing facility

D. Storage of pumpable waste

To avoid reactions of incompatible liquid/pasty materials, a larger number of storage tanks/systems need to be made available at the site. Such arrangement avoids the danger of explosion or polymerization. The design, material selection, and construction of the valves, pipelines, tanks, and seals need to be adapted to the waste characteristics. They must be corrosion-proof and offer the option of cleaning and sampling. It may be necessary to homogenize the tank contents with agitators. These agitators may be hydraulic or mechanical in design. Some tanks may be requiring heating depending on the waste characteristics. Figure 6.6 depicts the storage arrangement for liquid pumpable materials.

E. Safety aspects of storage

The following measures are desired in storage areas:

(a) Alarms to alert about emergency situations.
(b) Well-maintained and manned communications system at the site. This helps to have contact with the control room and the local fire department in case of any fire emergency.
(c) All electrical equipment need to be grounded with appropriate devices.

Fig. 6.6   Tanks for liquid hazardous wastes

F.   Fire detection and control systems

When storing flammable liquid waste, automatic fire control systems should be provided. These would include foam, water, and cannons with options to use foam or water, dry powder, and carbon dioxide. Nitrogen blanketing may be needed for pre-treatment and kiln loading of hazardous wastes.

Temperature measurements and monitoring should be installed whenever feasible. Temperature variations can be used to trigger alarms.

When ammonia is used, NH3 detection and water spray devices to absorb its releases need to be put up as a safety measure.

6.6.6   Best Available Techniques (BAT) and Best Environmental Practice (BEP)

Large amounts of natural materials get conserved when secondary resources are utilized. These include AFRs and cementitious materials such as fly ash, slags, and chemical gypsum.

A.   BAT/BEP for cement production

Dry preheater/pre-calciner kilns are considered as one of the best available techniques (BAT) which contribute to the best environmental practice (BEP). This is the most economically feasible technological option. BAT has a competitive advantage and contributes to gradually phasing out of outdated, polluting, and less competitive technologies.

Dry process kiln with multi-stage preheating and pre-calcination is the best available technique for the production of cement clinker for new plants and major upgrades. For reduced kiln emissions and efficient energy usage, the kiln needs to be run as a smooth and stable process and operating close to the process parameter setpoints.

Careful monitoring and control of the input substances entering the kiln help in the reduction of the emissions. When feasible, it is desired to select homogenous AFRs with low contents of sulphur, nitrogen, chlorine, metals, and volatile organic compounds.

B.   BAT/BEP for controlling emissions of PCDD/PCDFs

It is important to note that if alternative raw material includes elevated concentrations of organics, feeding it as a part of the raw material mix should be avoided. Further, no AFR should be fed during start-up and shutdown. Quick cooling of the kiln exhaust gases to lower than 200 °C is the most important measure to avoid PCDD/PCDF formation in wet kilns. This feature is already present in the process design of the modern preheater and pre-calciner kilns and has air pollution control device (APCD) operating at temperatures less than 150 °C.

C.   Conventional fuels

There are three different types of conventional fuels are used in cement kiln in decreasing order of importance. These are fossil in nature.

- Pulverized coal and petcoke.
- Fuel oil (heavy).
- Natural gas.

Cement kilns are normally operated at the lowest feasible excess oxygen levels to keep heat losses to the minimum. For easy and complete combustion of these fuels in the kiln system, fuel is fed in a uniformly processed form with reliable fuel metering systems. These conditions are essential for all types of natural and alternative fuels.

The different fuel feed points into the cement kiln system are the following.

- Main burner at the rotary kiln.
- Pre-calciner burners.
- Feed chute to the pre-calciner.
- Mid-kiln valve to long wet and dry kilns (for lump fuel).
- Fuel burners at the riser duct.
- Feed chute at the transition chamber at the rotary kiln inlet end.

The fuel fed in the main burner of the kiln produces the flame having temperatures around 2000 °C. The flow rate and pressure of the primary air help in adjusting the shape and thermal characteristics of the flame in the main burner.

6.6.7 Co-processing of AFRs

Co-processing of AFRs is required to be carried out in compliance with the permit issued by the authorities. These permits are issued in many ways.

- Some do not encourage co-processing of AFRs/hazardous wastes beyond certain categories or concentration limits.
- Some specify an explicit list of acceptable AFRs. This list may prescribe certain maximum and/or minimum values for parameters such as chlorine content, calorific value, and heavy metals.
- Some regulations specify a negative list with waste categories not allowed.
- Some focus on emissions limits only.

The waste categories that can be accepted for co-processing at the specific plant depend upon the local raw material and fuel chemistry, the availability of equipment for handling, feeding the waste materials, and controlling the same.

A. Input control

To maintain stable conditions during kiln operation, consistent long-term availability of appropriate wastes is required. Content of various elements of concern in them needs to be specified and controlled. These elements include metals, VOCs, fluorine, sulphur, chlorine, nitrogen, etc. Their limitations in the product and/or the process need to be established and defined. While feeding waste to the kiln, it is important to ensure that the waste materials are getting exposure to:

(a) Sufficient mixing conditions.
(b) Sufficient oxygen.
(c) Sufficient retention time.
(d) Sufficient temperature.

Determination of the appropriate feeding point for the waste depends upon the waste type and its composition. Wastes should not be fed as part of raw mix feed if it contains organics and not be fed during start-up and shutdown. Automated monitors should be employed to alert operators, for example, in the event of sudden pressure drop due to pipe rupture or pump failure, a pressure transducer should turn off the waste fuel pump automatically. The pressure transducer is located in the waste piping at the entrance of the kiln.

Interlocks should stop the flow of waste automatically while the normal fuel or feed supply and/or the combustion airflow is interrupted. Interlocks should also stop the flow when CO levels indicate a problem.

B. Selection of the feed point

A constant feed rate and quality of the waste material are essential to ensure that the use of AFRs does not destabilize the smooth operation of the kiln. This is also

Fig. 6.7 Liquid hazardous waste fed through the main burner

required to be done to ensure that the site's normal environmental performance or the product quality is not impacted. The selection of the feed point for wastes into the kiln depends upon the nature of the wastes used.

(a)   Only the main burner is required to be utilized for co-processing highly chlorinated organic compounds and persistent organic pollutants (POPs). This is to ensure that their destruction occurs to the desired level due to the long retention time and high combustion temperature available in the kiln. Other feed points are selected only when high levels of destruction and removal efficiency (DRE) are demonstrated through tests.

Figure 6.7 depicts the feeding arrangement of liquid AFR to the main burner.

(b)   AFRs with high volatile organic components should be fed into the high-temperature zones of the kiln system directly

Figure 6.8 depicts the solid feeding arrangement in the kiln inlet.

(c)   Mineral inorganic wastes that do not have organic constituents can be fed in the raw meal or raw slurry preparation system.

Figure 6.9 depicts the feeding arrangement of feeding the material to the raw mill.

(d)   Operations and process control

It is desired to specify acceptable operating limits of the retention time, oxygen levels, feed rates, temperatures, etc., for each waste. Acceptable composition and

Fig. 6.8 Feeding of solid waste to the kiln inlet

variations in the physical and chemical properties of the waste also need to be specified. Principles of good operational control need to be followed and all relevant process parameters need to be monitored and recorded. These include the following.

(a) Free lime.
(b) Oxygen concentration.
(c) Carbon monoxide concentration.

(e) Kiln operation and feeding of wastes

During co-processing, it is desired that the cement plant operates in a smooth and steady manner. It is desired to establish reference data by adding controlled doses of waste, evaluate the changes, and implement required practices to control emissions. The impact of wastes on the total input of circulating volatile elements such as chlorine, sulphur, or alkalis must be assessed carefully prior to acceptance of the waste to avoid operational troubles in the kiln system. Input limits and operational setpoints for these components shall be based on the site conditions.

In case of failure of any plant equipment, procedures for stopping waste feed must be designed and implemented. The logic for each of the feed cut-off considerations must be specified properly for clear understanding. Waste and AFR should only be fed when normal operating temperatures are achieved in the kiln system. The same should not be fed to the kiln during start-up, shutdown, or major kiln upset conditions.

(f) Laboratory and quality control

A well-equipped laboratory with appropriate facilities for sampling and testing is essential for successful co-processing. To monitor the performance and to improve

Fig. 6.9 Feeding of mineral inorganic wastes in the raw mill

it, inter-laboratory tests need to be carried out periodically. The laboratory personnel must be competent and trained in analysing different kinds of AFRs, hazardous, and non-hazardous wastes.

Wastes/AFRs, raw materials, and fuels entering the cement plant for co-processing or being produced in cement plant need to be controlled regularly. An appropriate QA/QC plan must be prepared and utilized. This should include the following.

(a)   Sampling.
(b)   Frequency of sampling and analysis.
(c)   Laboratory protocols and standards.
(d)   Recording and reporting protocol.
(e)   Calibration procedures and maintenance.
(f)   Personnel assignment.

Fig. 6.10 Sampling and analysis of AFRs

Fig. 6.11 Bomb calorimeter

Figures 6.10, 6.11, 6.12, 6.13, 6.14, 6.15, and 6.16 depict some of the practices and instruments associated with the QA/QC of the AFRs

6.6.8   Cement Quality

The most important aspect while undertaking co-processing is to ensure that the desired quality cement is produced. For this, evaluation of chemical and physical characteristics of all relevant parameters is a must. It should address all aspects concerning potential clinker contamination and cement quality and data must be kept recorded.

Fig. 6.12 Flash point
analyser

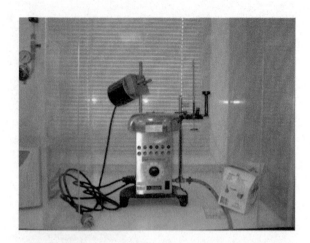

Fig. 6.13 Chloride titrator

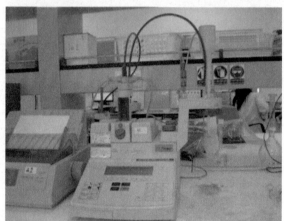

Fig. 6.14 Sulphur analyser

Fig. 6.15 GC–MS

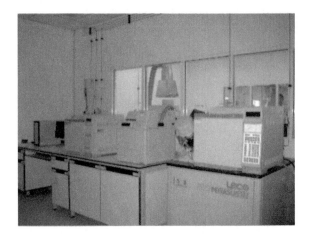

Fig. 6.16 ICP AES

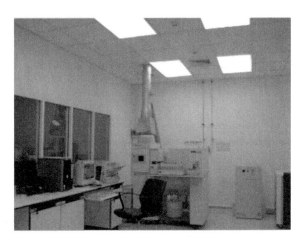

The fact that the co-processing of AFRs has not affected the cement quality must be documented. The following needs to be considered:

(a)   Zinc, Phosphate, and Fluorine influences the strength and setting time of the cement.
(b)   Alkalis, Sulphur, and Chlorine contents in waste / AFR keep circulating in the kiln system and impact the overall product quality.
(c)   For sensitive users, allergic reactions may be caused by Chromium.

Cements are normally tested in terms of their strength with standard test procedures.

6.6.9   Emission Monitoring

To demonstrate compliance with existing regulations and agreements, emission monitoring is obligatory. Controls are necessary on the input of conventional materials and their potential impacts. Organic carbon in raw materials results in volatile organic compound (VOC) emissions, CO, and CO2; Sulphides in raw materials result in the release of SO2. Heavy metals in fuel and raw material need to be assessed, monitored, and controlled. This is true especially for the heavy metals that are volatile in nature because they are not completely captured in the clinker.

A.   Emission limit values

Directive 2000/76/EC governs the co-processing AFR and treating hazardous wastes in the cement kilns in the EU. The emissions limits in flue gases given in table 1 below, corrected to 273 K, 101.3 kPa, 10% O2 and dry gas (IPCC, 2013; Council Directive, 2000) are to be complied with in the EU. Daily average values of pollutants for cement plants co-incineration of hazardous waste are given in table 6.1. In co-incineration of hazardous waste less than 40% of the resulting heat release must come from waste at 10% O2, dry gas.

B.   Continuous monitoring

Continuous emission monitoring system (CEMS) should be in place in the exit stack to monitor the following parameters online:

(a)   Exhaust volume.
(b)   Humidity.
(c)   Temperatures.

Table 6.2 Co-incineration of hazardous waste: Daily average values of pollutants in cement plants

Pollutant	C (all values in mg/m3)
Total dust	30
HCl	10
HF	1
HO_x	500[1]/800[2]
Cd + Tl	0.05
Hg	0.05
Sb, As, Pb, Cr, Co, Cu, Mn, Ni, V	0.5
Dioxins and furans	0.1 ng TEQ/m3
SO_2	50[3]
TOC	10[3]

(1) new plants (2) existing plants (3) exceptions may be authorized by the component authority in cases where SO2 and TOC do not result from the waste

(d) Particulate matter.
(e) O_2.
(f) NO_x.
(g) SO_2.
(h) CO.
(i) Volatile organic compounds (VOC).
(j) HCl.
(k) Pressure.

C. Regular monitoring

Some parameters should be monitored on a periodical basis in the exit stack:

(a) Metals and their compounds.
(b) Chlorobenzenes, HCB, and PCBs including coplanar congeners and chloro-naphthalenes.
(c) Total organic carbon.
(d) HF.
(e) NH_3.
(f) PCDD/PCDF.

D. Occasional monitoring

Occasional monitoring is required under special operating conditions for the following parameters (Karstensen, 2008):
    Figure 6.17 depicts the photo of the emission monitoring being carried on the cement kiln stack.

(a) Benzene, toluene, and xylene.
(b) Other organic pollutants.
(c) Polycyclic aromatic hydrocarbons.
(d) Destruction and removal efficiency (DRE).
(e) Destruction efficiency (DE).
(f) Heavy metals such as Cd, Tl, Hg, Sb, As, Pb, Cr, Co, Cu, Mn, Ni, and V.

E. Additional measures for exit gas cleaning

Pollutants such as ammonia (NH_3), ammonium (NH_4^+) compounds, hydrogen chloride (HCl), hydrogen fluoride (HF), organic compounds, sulphur dioxide (SO_2), heavy metals, and residual dust are usually removed from the exhaust gas in conventional filters. An activated carbon filter has high removal efficiency for trace pollutants such as mercury and PCDD/PCDF (>90%). Selective catalytic reduction (SCR) can be applied for reducing NOx to the desired extent from the stack gases. The efficiency of NOX removal is very high in the SCR process. In this process, NO and NO2 get converted to N2 by reacting it with NH3 over a catalyst at a temperature below 400 °C.

Fig. 6.17   Stack gas emission monitoring

6.6.10   Test Burn and Performance Verification

Test burns are required to be carried out in cement kilns to demonstrate the destruction of principal organic hazardous compounds (POHC) through destruction and removal efficiency (DRE) and destruction efficiency (DE) (Karstensen, 2011, 2014; Karstensen et al., 2006, 2010). Test burns with hazardous compounds require independent verification and professional supervision.

The DRE considers emissions to air only. The DE considers emissions to all three streams, namely solid, liquid, and gas. It is the most comprehensive way of verifying the co-processing performance.

The following conditions should be fulfilled in a test burn:

(a)   For POPs, the DRE/DE should be $\geq 99.9999\%$. The DRE for other hazardous compounds should be $> 99.99\%$.
(b)   PCDDs/PCDFs emissions limit of 0.1 ng TEQ/Nm3 needs to be demonstrated both under baseline and test burn conditions.
(c)   Parallelly, existing emission limit values need to be complied with by the cement kiln.

Test burns with non-hazardous waste are done to evaluate their impact on the process and clinker product. These simplified tests are usually carried out by process engineers at the cement plant. In these tests, already installed online monitoring equipment is utilized to generate the required data.

References

Basel Convention. (2007). General technical guidelines for the environmentally sound management of wastes consisting of, containing or contaminated with persistent organic pollutants (POPs). http://www.basel.int/techmatters/techguid/frsetmain.php?topicId=0.

Council Directive. (2000). Council Directive 2000/76/EC on the Incineration of Waste. *Official Journal of the European Communities, Brussels, Official Journal L 332,* 28/12/2000.

Dahai Yan, Zheng Peng, Kåre Helge Karstensen, Qiong Ding, Kaixiang Wang, & Zuguang Wang. (2014). Destruction of DDT wastes in two preheater/precalciner cement kilns in China. *Science of the Total Environment, 476–477*(2014), 250–257. ISSN No. 0048–9697.

GTZ-Holcim. (2006). Guidelines on Co-Processing Waste Materials in Cement Production. http://www.holcim.com.

IPPC. (2013). Best Available Techniques (BAT) Reference Document for the Production of Cement, Lime and Magnesium Oxide: Industrial Emissions Directive 2010/75/EU. 978-92-79-32944-9.

Kåre Helge Karstensen, Ulhas V. Parlikar, Deepak Ahuja, Shiv Sharma, Moumita A. Chakraborty, Harivansh Prasad Maurya, Mrinal Mallik, Gupta, P. K., Kamyotra, J. S., Bala, S. S., & Kapadia, B. V. (2014). Destruction of concentrated Chlorofluorocarbons in India demonstrates an effective option to simultaneously curb climate change and ozone depletion. *Environmental Science and Policy, 38*(2014), 237–244. ISSN No. 1462–9011.

Karstensen, K. H. (2014). Destruction of hazardous chemicals and POPs in cement kilns. Zement, Kalk und Gips—ZKG International, July 2014. ISSN: 0949–0205.

Karstensen, K. H. (2011). Compilation of performance verification and trial burns results in cement kilns. Chapter in Basel Convention Technical guidelines on the environmentally sound co-processing of hazardous wastes in cement kilns. UNEP, Geneva, 11 November 2011. http://www.basel.int/TheConvention/Publications/BrochuresLeaflets/tabid/2365/Default.aspx. ISBN No. UNEP/CHW.10/6/Add.3/Rev.1.

Karstensen, K. H., Mubarak, A. M., Bandula, X., Gunadasa, H. N., & Ratnayake, N. (2010). Test burn with PCB in a local cement kiln in Sri Lanka. *Chemosphere, 78,* 717–723. ISSN: 0045–6535.

Karstensen, K.H. (2008). Formation, release and control of dioxins in cement kilns—A review. *Chemosphere, 70,* 543–560.

Karstensen, K. H., Kinh, N. K., Thangc, L. B., Viet, P. H., Tuan, N. D., Toi, D. T., Hung, N. H., Quan, T. M., Hanh, L. D., & Thang. D. H. (2006, October). Environmentally sound destruction of obsolete pesticides in developing countries using cement kilns. *Environmental Science & Policy, 9*(6), 577–586. ISSN No. 1462–9011.

Karstensen, K. H. (2006). Formation and Release of POPs in the Cement Industry. Report to the World Business Council for Sustainable Development. 30 January.

UNEP. (2007). Stockholm Convention Expert Group on Best Available Techniques and Best Environmental Practices. Expert group on BAT/BEP - Cement Kilns firing hazardous Waste, submitted February 2007. UNEP. http://www.pops.int/documents/guidance/batbep/batbepguide_en.pdf.

World Business Council for Sustainable Development (WBCSD). (2006). Guidelines for the Selection and Use of Fuels and Raw Materials in the Cement Manufacturing Process. www.wbcsd.com.

Chapter 7
Sustainability Considerations in Cement Manufacturing and Co-processing

7.1 Introduction

Cement manufacturing is a highly resource and energy-intensive process. In addition, it also emits a significant amount of CO_2 due to calcination of limestone, raw material, and also due to firing of fossil fuels. Cement industry has a large carbon footprint accounting for over 7% of the CO_2 released globally. The same figure for India also works out to about 7%. Cement industry also contributes to pollution in the environment due to the release of emissions such as Particulate Matter, SOx, NOx, VOC, etc. Cement industry also undertakes limestone mining which impacts the local ecosystem and biodiversity. The Cement industry is not water intensive, but water consumption is certainly a critical resource during the manufacturing process. It utilizes a considerable amount of plastics as the packaging material for cement. Thus, irreversible use of natural resources, large carbon footprint, release of air pollutants, biodiversity impacts, water consumption, and plastic packaging are the critical concerns of the industry from a sustainability point of view. To be sustainable, cement industry is undertaking several initiatives to address these concerns: Environmental, Economic, and Social. Following measures are being taken up as sustainability-related initiatives by the corporates. (1) Reducing CO_2 footprint, (2) Waste reduction, (3) Investing in renewable energy, and (4) Helping agencies that are supporting the cause of sustainability. They are also making bold announcements with respect to their journey towards sustainability.

7.2 Sustainable Development Goals (SDG)

The Sustainable Development Goals (SDGs) were adopted by the United Nations in 2015 as a universal call to take action to end poverty, protect the planet and ensure that all people enjoy peace and prosperity by 2030 (UNDP, 2015). These goals are 17 in number as mentioned in Fig. 7.1.

© The Author(s), under exclusive license to Springer Nature Singapore Pte Ltd. 2022
S. K. Ghosh et al., *Sustainable Management of Wastes Through Co-processing*,
https://doi.org/10.1007/978-981-16-6073-3_7

Fig. 7.1 United nations sustainable development goals

A total of 193 countries have adopted this 2030 agenda and are leading actions on the same through a well-defined action plan and mechanism. The Indian Cement Sector also has formulated its SDG Roadmap and has launched in 2019.

7.3   Cement Sustainability Initiative (CSI)

Twenty-four major cement producers, representing about 30% of the cement production of the world, with operations in more than 100 countries have come together to take up the Cement Sustainability Initiative (CSI). CSI members consider that there is a strong business case to pursue the agenda of sustainable development.

CSI is a part of the World Business Council for Sustainable Development (WBCSD). It has focussed on understanding, managing, and minimizing the impacts of cement production and use by addressing different issues. In 1999, under the auspices of the WBCSD, 10 leading cement companies commissioned the Battelle Memorial Institute, a US-based not-for-profit consulting firm, to conduct independent research into how the cement industry could meet these sustainability challenges. Battelle's final report, Towards a Sustainable Cement Industry, was released in April 2002. (CSI, 2002). The companies responded to the Battelle recommendations by issuing an Agenda for Action, outlining individual and joint actions by the industry in adopting the sustainability initiatives.

The major initiative of the cement industry is reducing its carbon footprint. The other parameters include biodiversity, water conservation, plastics reduction, improving resource efficiency, conserving natural resources, reducing emissions, etc.

In 2019, the work carried out by the Cement Sustainability Initiative (CSI) is officially transferred from the World Business Council for Sustainable Development (WBCSD) to the Global Cement and Concrete Association (GCCA). The Global

Cement and Concrete Association (GCCA) was formed in January 2018 by seven major cement companies, including several WBCSD CSI members.

7.4 CSI and Low-Carbon Technology Road (LCTR) Map

Reducing the carbon footprint in the cement manufacturing process is an important requirement of the cement industry. Drawing a road map for reducing carbon footprint in the cement industry is an important requirement. Indian Cement Industry has prepared this road map in the year 2013. International Energy Association and Cement Sustainable Initiative have together brought out a report Technology Roadmap Low-Carbon Transition in the Cement Industry (IEA, 2013). These road maps define the action plans to achieve the desired level of carbon emission reduction.

7.5 GCCA and Getting Numbers Right

Global cement production continues to rise due to growing demand. Estimates suggest that 4.1 Giga Tonnes of cement was produced globally in 2019. China is the largest cement producer, accounting for about 55% of global production, followed by India at 8%. Global Cement and Concrete Association (GCCA) represents about 30% of the total cement production of the world. GCCA is guiding the implementation of sustainability initiatives of cement and concrete manufacturing industries worldwide. It also monitors and documents the relevant information related to the sustainability initiatives of the cement and concrete sector.

It is compiling a database called "Getting Numbers Right" (GNR). In this database, key sustainability-related data pertaining to the cement industry is monitored and recorded (GCCA, 2019). This includes data pertaining to cement production, CO_2 emissions, Power production and Consumption, Heat production and Consumption, co-processing of Alternative Fuels and Raw materials, and Mineral Components. Table 7.1 provides data pertaining to the production of clinker and cement by the GCCA member cement companies.

The data tabulated in this chapter pertains to these production figures which can be seen as increasing year on year. To document the global status on co-processing, the data available in the GNR database of GCCA is utilized. This data does not

Table 7.1 Production of Clinker & Cement by member companies of GCCA

Material	Year	1990	2000	2006	2012	2018
	Units					
Clinker	Million TPA	423	520	626	642	650
Cement	Million TPA	502	628	797	852	861

represent the total cement industry of the world but represents a sizable portion of the same to provide a representative view of the sustainable growth pursued by the cement industry. This database is updated till 2018.

Following data from the GNR database is represented as graphs to represent the sustainability initiatives of the cement industry:

a. Clinker Factor,
b. TSR% due to use of AFRs,
c. CO_2 released per tonne cement,
d. Thermal Energy Utilized per Ton of Clinker,
e. Electrical Energy utilized per Ton of Cement.

Although the data is documented by GCCA on a yearly basis, for representation purposes, five-yearly data is utilized in these graphs.

7.6   Carbon Footprint

The CO_2 emissions from the cement industry consist of Scope 1 emission, Scope 2 emission, and Scope 3 emissions. The different levers which are identified in the carbon footprint reduction road map are as follows;

A. AFR co-processing,
B. Energy efficiency improvement,
C. Clinker substitution,
D. Waste Heat Recovery, and
E. New technologies that are under development.

7.6.1   Scope 1 Emissions

These are the CO_2 emissions occurring directly from the cement manufacturing operations. Almost half of the CO_2 emissions come from the calcination of limestone in the raw mix when calcium carbonate is thermally decomposed, producing lime and carbon dioxide. A significant proportion of CO_2 is emitted during the combustion of fossil fuels.

International Energy Agency (IEA) along with World Business Council for Sustainable Development (WBCSD) have teamed together to identify these avenues and have brought out specific approaches on the same. Since 2002, cement-producing companies in the Cement Sustainability Initiative (CSI) have initiated measuring, reporting, and mitigating their CO_2.

This initiative has made considerable progress in mitigating CO_2 levels. In 2009, the urgent need of identifying technology to reduce the energy use and CO_2 intensity in cement production was felt and CSI member companies around the world worked with the IEA on this subject. Through this collaboration, the first industry roadmap

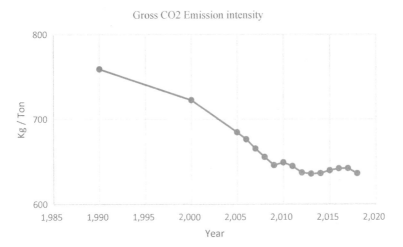

Fig. 7.2 Reduction in Primary CO_2 Emissions from the cement industry

to mitigate the CO_2 emissions was worked out. That roadmap defines emissions reduction potential from such technologies that can be implemented in the cement industry.

A similar road map was also prepared for Indian Cement Industry considering its large capacity of cement manufacture in the country and was published in 2013. (IEA and WBCSD, 2013). Figure 7.2 provides the status of Primary CO_2 emission reduction achieved by the global cement industry as reported in the GNR by GCCA.

A. *Alternative fuels and raw materials:*

By using this pillar, CO_2 emission reduction is achieved by promoting the use of wastes as Alternative Fuels and Raw materials. The wastes, that are combustible in nature, such as sorted municipal waste, industrial wastes, and biomass get utilized as Alternative Fuels. They replace the fossil fuels. These wastes would otherwise get combusted in incinerators or get land-filled or improperly dumped. The CO_2 footprint of biomass is zero and it is considered carbon neutral. The CO_2 released from other waste streams has a lower carbon footprint than fossil fuels.

There are also waste materials that can replace other mineral elements such as Calcium, Iron, Aluminium, and Silica used in cement manufacture. These include lime sludges from the paper industry, ETP sludge from the chemical industry, and water treatment plants. These provide Calcium elements in cement manufacture.

Iron sludges, Mill scale, Iron scrap, etc., provide Iron, Aluminium dross provides Aluminium while Aluminium and Iron are provided by Red mud and silica is obtained from rice husk ash. These waste streams replace the natural materials that are utilized in cement manufacture to derive these elements. The advantage of AFR utilization is that it facilitates reduction in CO_2 emissions and also facilitates reduction in the use of mined minerals.

Figures 7.3, 7.4, 7.5, 7.6 provide the data pertaining to AFR utilization in the GCCA member cement industries

These figures clearly demonstrate the sustainability aspiration of the cement industry located worldwide. These figures also indicate that the sustainability agenda requires time for implementation.

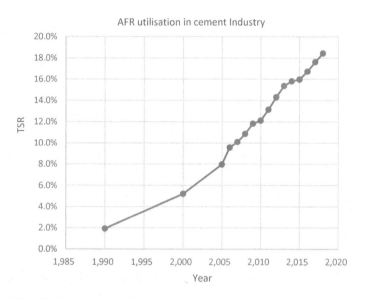

Fig. 7.3 AFR utilized in the cement industry

Fig. 7.4 Use of different fossil Fuels by the cement industry

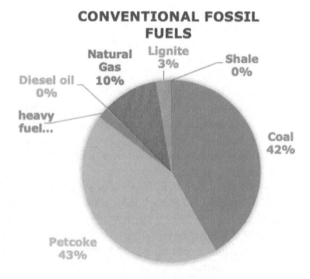

Fig. 7.5 Use of different Alternative Fuels by the cement industry

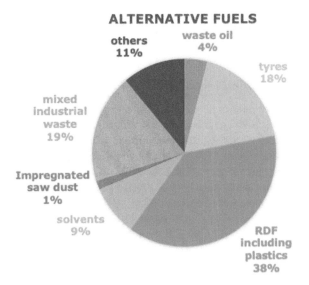

Fig. 7.6 Use of different Biomass Fuels by the cement manufacture

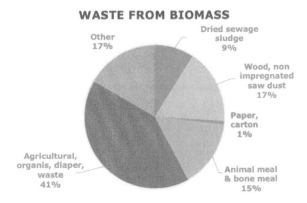

In the given 28 years of the monitored data (GCCA, 2019), one can observe that TSR improved from 2 to 18%, Clinker Factor reduced from 84.3 to 75.4%, CO_2 emissions reduced from 776 kg / T Cement to 642 kg / T Cement. Electrical Energy reduced from 119 KWH / T cement to 102 KWH / T Cement and the specific energy consumption reduced from 4250 GJ / T Cl to 3483 GJ / T Cl. It needs to be noted here that all these figures are average figures of all the representative plants.

B. *Thermal and electrical energy efficiency:*

By deploying state-of-the-art technologies in new cement plants, and retrofitting more energy-efficient equipment in existing plants the CO_2 emissions get reduced. The other initiatives include reducing heat losses from the preheater, kiln systems and cooler sections, reducing the ambient air ingress through the joints in the equipment and systems, increasing the heat transfer efficiency of the kiln and cooler sections by providing additional heat transfer stages or areas, etc., also has helped in improving thermal efficiency and reducing CO_2 emissions.

Figures 7.7 and 7.8 depict the improvement the cement industry has achieved in the electricity and thermal energy usage in the manufacture of cement, respectively.

C. *Clinker substitution:*

Pozzolanic materials namely, (i) Fly ash, (ii) Granulated blast Furnace Slag, (iii) Natural volcanic ash, etc. are utilised as substitution to clinker in the cement manufacture. This substitution helps in avoiding the CO2 released in clinker manufacturing. Figure 7.9 provides the improvement achieved in the clinker factor by cement industries. Since clinker gets substituted with these materials, all the CO_2 released in its manufacture gets avoided. These pozzolanic materials include (i) Fly ash, (ii) Granulated blast Furnace Slag, (iii) Natural volcanic ash, etc. Figure 7.9 provides the improvement achieved in the clinker factor.

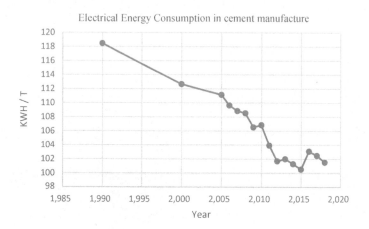

Fig. 7.7 Electrical energy consumption in cement manufacture

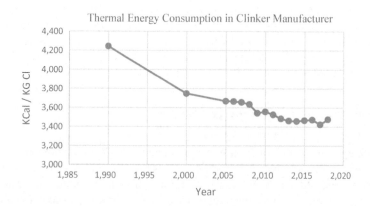

Fig. 7.8 Thermal energy consumption in clinker manufacturer

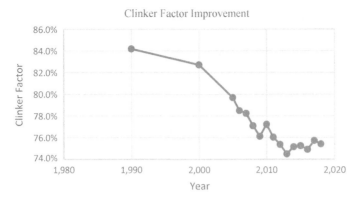

Fig. 7.9 Clinker factor improvement achieved by the cement industry

D. *Newer technologies:*

Newer technologies that could substantially improve the CO_2 reduction potential of the cement industry are the following:

i.   Use of mineralizers,
ii.   Geopolymer cement and use of nanotechnology in cement production,
iii. Fluidized-bed advanced cement kiln system (FAKS),
iv.   Carbon used for algal growth for biofuels production,
v.   Replacing thermal energy with electrical energy coming from renewable sources,
vi.   Carbon capture and storage.

7.6.2 Scope 2 Emissions

This carbon is on account of the use of electrical energy derived from the local or external thermal power plant. This can be reduced using the following technological interventions:

A. *Use of renewable electrical energy for operating cement plant,*
B. *Avoiding the use of electricity in the cement manufacture that is produced using fossil fuels,*
C. *Waste heat recovery in the cement manufacturing.*

  Waste Heat Recovery technology is used to convert the excess thermal energy available in the cement kiln into electricity. This partially offsets the electrical energy requirement in the cement manufacturing process. The waste heat recovery is feasible from the pre-heater system as well as from the cooler system. Generally, the acid due point of the exhaust gases determines the lowest temperature to which the systems are designed. Following different technology-based WHR designs are available for implementation in the cement plants.

(i)   **Conventional Rankine cycle:**

This design utilizes steam as the power transfer media that is utilized in conventional Thermal Power Plants. Here the temperature of the hot gas needs to be above 350 OC.

(ii)   **Organic Rankine Cycle:**

This design utilizes an organic compound (Pentane) as the power transfer media instead of steam used in the conventional Rankine cycle. Here the temperature of the hot gas may be lower than 350 OC also.

(iii)   **Kalina Cycle:**

This design utilizes a mixture of Ammonia and water as the power transfer media instead of steam used in the conventional Rankine cycle. Here the temperature of the hot gas may be lower than 350 OC also.

7.6.3   Scope 3 Emissions

The scope 3 carbon footprint is on account of transportation of the cement production in the marketplace. This can be reduced by reducing (a) the average distance of transportation by concentrating on the local markets (b) by using biofuels for transportation.

7.7   Conservation of Natural Resources and Circular Economy

The major release of CO_2 happens from the calcination of limestone. By using already calcined lime bearing wastes in the cement manufacture, the limestone usage gets reduced and hence the CO_2 emissions. There are also some new technological innovations that are awaiting commercialization such as LC3 cement. LC3 is a new type of cement that is based on a blend of limestone and calcined clay (LC3). These cements have substantially low carbon footprint than the conventional cement such as OPC.

Out of the different levers elaborated above, co-processing is one of the major levers available for the cement industry. Technically, it is possible to replace large amount of fossil raw materials and fossil fuels using wastes, thereby promoting circular economy. There are a few operating cement plants globally that have been able to replace fossil fuels to an extent of >90%. Hence, almost all cement plants have aspiration to improve the AFR utilization in their cement plants so as to address the sustainability factors.

7.8 Reduction in Emissions

The major emission concerns from cement plants are NOx emissions generated during thermal treatment and SOx emissions—if any, from the use of sulphur-containing raw materials. The SOx can be converted into gypsum using the FGD technology. In case we want to treat both SOx and NOx emissions, then the same can be converted into Ammonium Sulphate and Ammonium Nitrate fertilizer. This technology development has been done in Japan. Commercial application of this process has been achieved, with the co-operation of the Chinese government and EBARA Corporation in 1997 at the coal-fired Chengdu Power Station in China (Yoshitaka Doi, 2000).

7.9 Water Conservation and Harvesting

Many efforts are being made by the cement industry to become water positive by harvesting the rainwater in the old exhausted mine sites. There are other benefits of water harvesting (a) improvement of the water table in the surrounding region, (b) availability of water during the summer season, (c) improving greenery in the plant/mines/surrounding area due to availability of water for plantation, (d) improving the agricultural productivity, etc.

If the cement plant's efforts help in conserving more water than the one utilized by it in the manufacturing operations, then it becomes water positive. Many cement plants these days are putting efforts to become water positive.

7.10 Plastic Packaging

Many cement companies package the cement in plastic bags and then send it to the marketplace. After the intended use of this packaging is over, it's utilized in some other applications such as material filling at the construction site, rain protection, etc. Subsequently, it becomes a waste material adding to the plastic waste management problem. There are two options available for the cement plants to tackle this problem: (a) To avoid packaging in plastic bags and resort to bulker-based movement. (b) Co-process plastic waste received from the marketplace. This way if the cement plant can co-process more plastic waste than the plastic bag weight employed by it in packaging, then the company becomes plastic negative.

7.11   Biodiversity

Mining activity in the cement industry involves blasting which causes sound pollution and vibrations. Vibrations displace the soil and cause vegetation loss. On account of these effects, a lot of biological species get directly or indirectly impacted. Several initiatives are being employed by the cement industry to mitigate the impact on local biodiversity and the ecosystem. These include the following: (a) Sustainable mining practices to protect and enhance the landscape and biodiversity value of the area around the mines such as surface mining, controlled blasting to minimize dust and noise, covered transportation of raw material, development of water bodies and pastureland, plantation of native species, and land rehabilitation, etc. Setting up nesting and breeding habitats for migratory and local avifauna is another practice that the industry employs. Green belts in and around the mine lease and plant areas are also developed.

7.12   Challenges Faced in Implementing Sustainability Initiatives

The cement industry is pursuing a sustainability agenda with different sets of parameters. There are, therefore, different learnings with the cement industry with respect to pursuing different sustainability parameters. While undertaking the sustainability journey, several challenges are faced by the cement industry. These are discussed below.

A.   Resistance to change:

Pursuing the sustainability journey requires implementing changes in the existing business processes. What is grey will sell is the general tendency in the cement industry, and therefore, a huge resistance is encountered in implementing the changes. To achieve the desired changes in a successful manner, management commitment is the most important aspect. It also requires a dedicated, task-oriented, and passionate team to implement the same.

B.   Desired Technology Interventions:

The sustainability journey requires technology interventions and the availability of the same is an important aspect. Co-processing, clinker factor reduction, CO_2 capture and storage, reduction in energy consumption, etc. certainly require technology intervention. These technologies are new and need assimilation and adaptation.

C.   Cost of implementing technology:

All the sustainability-related technologies need additional capex and opex. These need to be factored into the budgets beforehand. This requires special efforts in its planning and implementation.

D.   **Capacity building:**

All these technologies also need to be appropriately digested into the routine operational processes for their effective implementation. This requires capacity building in the operating team through offline and online training sessions.

E.   **Scaling the learning curve:**

While implementing each of the sustainability initiatives, one needs to go through the learning curve. For example, to implement co-processing of a waste stream, one has to go through a large amount of learning in respect of its health and safety aspects, impact on the environment, process, quality of output, etc. This can be achieved through a gradual and progressive approach.

7.13   Sustainability and Co-processing

Co-processing is an important pillar of sustainability. It provides a huge opportunity to reduce GHG emissions and also reduce the environmental damage being caused by the wastes. It is a zero-waste technology and helps build a new circular economy. It also helps in replacing the natural resources with wastes and conserves them. It creates new opportunities for business. It supports ten SDGs, namely: (1) No poverty, (3) Good health and well-being, (6) clean water and sanitation, (9) industry, innovation, and infrastructure, (11) sustainable cities and communities, (12) responsible consumption and production, (13) Climate action, (14) life below water (15) Life on land (17) partnership for the goals.

7.14   Conclusions

Cement industry is a substantially sensitive industry to the cause of sustainability. It has undertaken a large number of sustainability initiatives as a voluntary measure that encompasses the entire gamut of manufacturing activities starting from quarry to lorry. Cement industry also has adapted to relevant BAT and BET options and has aligned itself to the SDGs defined by United Nations. Currently, the entire cement industry worldwide is adopting co-processing technology and is making reasonable progress in the same. One of the important advantages of co-processing is that it can adapt to the growing waste management needs of the society and facilitate circular economy.

References

CSI, Agenda for Action, July 2002. Retrieved from https://www.wbcsd.org/Sector-Projects/Cement-Sustainability-Initiative/Cement-Sustainability-Initiative-CSI.

GCCA, Getting the Numbers Right, 2019. Retrieved from https://gccassociation.org/sustainability-innovation/gnr-gcca-in-numbers/.

IEA, Technology Roadmap—Low-Carbon Technology for the Indian Cement Industry—Analysis—IEA Feb 2013. Retrieved from https://www.wbcsd.org/Sector-Projects/Cement-Sustainability-Initiative/Resources/Technology-Roadmap-Low-Carbon-Technology-for-the-Indian-Cement-Industry.

Michel Grant, Investopadia, Sustainability Definition updated October 13, 2020. Retrieved from https://www.investopedia.com/terms/s/sustainability.asp#:~:text=Sustainability%20focuses%20on%20meeting%20the%20needs%20of%20the,social%E2%80%94also%20known%20informally%20as%20profits%2C%20planet%2C%20and%20people.

UNDP, Sustainable Development Goals, 2015 Retrieved form https://www.undp.org/sustainable-development-goals#:~:text=The%20Sustainable%20Development%20Goals%20%28SDGs%29%2C%20also%20known%20as,by%202030%20all%20people%20enjoy%20peace%20and%20prosperity.p.org

Yoshitaka Doi et al. (2000, March). Operational experience of a commercial scale plant of electron beam purification of flue gas. *Radiation Physics and Chemistry, 57*(3–6), 495–499

Chapter 8
Waste Management Rules in India and Other Countries Focussing on Co-processing

8.1 Introduction

The environment comprises all entities, natural or manmade, external to oneself and their interrelationships which provide value, now or perhaps in the future, to humankind. Environmental concerns relate to their degradation through the actions of humans. Human beings carry out several different activities during which different kinds of wastes get generated. Generally, waste generators tend to discard them into the environment causing environmental degradation. This environmental degradation happens due to the pollution impact of the wastes on the air, soil, and water.

To protect the environment from such pollution, the governments of different countries enact different environment policies that consist of acts, laws, rules, guidelines, advisories, etc. In these policy frameworks, several technology options are also normally recommended and described along with the methodology of implementing them. Such policy-related information is provided in this chapter. This chapter explains the policy framework of India and some other countries with specific reference to the co-processing option for the management of wastes.

8.2 Environmental Regulation and Legal Framework in India—Constitutional Perspective

- Under Article 48-A of the constitution of India, the responsibility regarding environmental protection has been laid on the states. This article reads as follows:
  "The State shall endeavour to protect and improve the environment and to safeguard the forests and wildlife of the country".

- Under Article 51-A(g) of the constitution, Environmental protection has been defined as a fundamental duty of every citizen of this country. This article reads as follows:

© The Author(s), under exclusive license to Springer Nature Singapore Pte Ltd. 2022 157
S. K. Ghosh et al., *Sustainable Management of Wastes Through Co-processing*,
https://doi.org/10.1007/978-981-16-6073-3_8

"It shall be the duty of every citizen of India to protect and improve the natural environment including forests, lakes, rivers and wildlife and to have compassion for living creatures."

- Article 21 of the Constitution is a fundamental right which reads as follows:
  "No person shall be deprived of his life or personal liberty except according to procedure established by law."

- Article 48-A of the Constitution comes under Directive Principles of State Policy and Article 51-A(g) of the Constitution comes under Fundamental Duties.
- Under Article 47 of the Constitution, the states have been given responsibility regarding raising the level of nutrition and the standard of living to improve public health. This article reads as follows:
  "The State shall regard the raising of the level of nutrition and the standard of living of its people and the improvement of public health as among its primary duties and, in particular, the State shall endeavour to bring about prohibition of the consumption except for medicinal purposes of intoxicating drinks and of drugs which are injurious to health."
  The 42nd amendment to the Constitution was brought about in the year 1974. As per this amendment, the responsibility of the State Government is to protect and improve the environment and to safeguard the forests and wildlife of the country. Further, it is the fundamental duty of every citizen to protect and improve the natural environment including forests, lakes, rivers, and wildlife, and to have compassion for living creatures. Figure 8.1 demonstrates the environmental and associated legislation in India.

- As conferred by Article 246(1), while the Union is supreme to make any law over the subjects enumerated in List I, the States, under Article 246 (3), enjoy competence to legislate on the entries contained in List II, and both the Union and the States under Article 246(2) have concurrent jurisdiction on entries contained in List III. In the event of a clash, the Union enjoys a primacy over States in that its legislation in the Union and the Concurrent List prevails over State legislations. Also, the Parliament has residuary powers to legislate on any matter not covered in the three Lists (Art. 248). These are provided in Tables 8.1, 8.2 and 8.3.

8.3   Indian Legislative Framework Related to Environmental Protection

Many environmental-related policies have been enacted by the government of India after attaining independence in 1947. These include acts and rules. Acts are the overarching policy instruments that define the objectives and the framework in each act, there are several rules notified to meet the objectives of the act. To explain the genesis of rules in an illustrative manner, several guidelines are prepared and published.

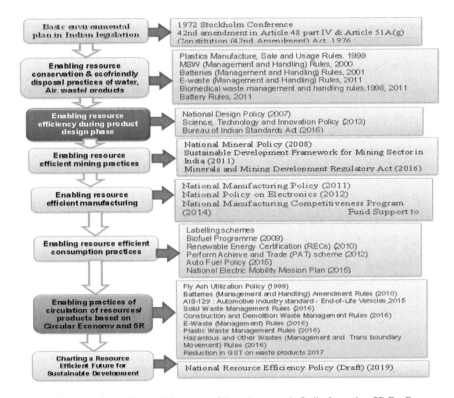

Fig. 8.1 Framework and key policies across lifecycle stages in India focussing SDGs, Resource conservation Circular Economy and 5R. (*Source* Ghosh, 2020)

Table 8.1 List I—union list

Entries	
52	Industries
53	Regulation and development of oil fields and mineral oil resources
54	Regulation of mines and mineral development
56	Regulation and development of inter-State rivers and river valleys
57	Fishing and fisheries beyond territorial waters

Table 8.2 List II—state list

Entries	
6	Public health and sanitation
14	Agriculture, protection against pest, and prevention of plant diseases
18	Land, colonization, etc
21	Fisheries
23	Fishing and fisheries beyond territorial waters
24	Fisheries

Table 8.3 List III—Common or Concurrent List

Entries	
17A	Forests
17B	Protection of wild animals and birds
20	Economic and social planning
20A	Population control and family planning

8.4   Important Acts Related to Environment and Co-processing

Following are the important environmental protection acts enacted by the government after the independence was achieved by India in 1947:

a.   The Factories Act of 1948.
b.   The water Act, 1974.
c.   The Forest (Conservation) Act, 1980.
d.   The Air Act, 1981.
e.   The Environment Protection Act, 1986.
f.   Motor Vehicles Act, 1988.
g.   The Public Liability Insurance Act, 1991.
h.   The National Environment Tribunals Act, 1995.
i.   The National Environment Appellate Authority Act, 1997.

All the above environment-related acts are important, and the objectives defined therein need to be addressed appropriately while undertaking the industrial activities.

The salient features of each of the above are explained in brief below.

8.4.1   The Indian Factories Act 1948

The Factories Act, 1948 (Act No. 63 of 1948), was amended by the Factories (Amendment) Act, 1987 (Act 20 of 1987). It serves to assist in formulating national policies in India in regard to occupational safety and health in factories in India. It deals with various issues related to safety, health, efficiency, and well-being of the persons at workplaces.

This Act is administered by the Ministry of Labour and Employment in India. This ministry is assisted by Directorate General Factory Advice Service and Labour Institutes (DGFASLI) and by the State Governments through their factory inspectorates. DGFASLI advises the Central and State Governments on the administration of the Factories Act and coordinating the factory inspection services in the States.

The Act is applicable to all factories using power and employing 10 or more workers. If the factory is not using power, then the same is applicable if it is employing 20 or more workers on any day of the preceding twelve months and in any part of which a manufacturing process is being carried on with the aid of power. This does not include a mine, or a mobile unit belonging to the armed forces of the union, a railway running shed or a hotel, restaurant or eating place (The factories Act, GOI, 1948).

8.4.2 The Water Act 1974

The Water Act represented India's first attempts to comprehensively deal with environmental issues. The Act prohibits the discharge of pollutants into water bodies beyond a given standard and lays down penalties for non-compliance. The Act was amended in 1988 to conform closely to the provisions of the EPA (Environment Protection Act), 1986. CPCB (Central Pollution Control Board) was established to develop standards and procedures for the prevention and control of water pollution. At the state level, the SPCBs (State Pollution Control Board) function under the direction of the CPCB and the state government (The Water Act, GOI, 1974).

8.4.3 The Forest (Conservation) Act, 1980

This Forest (Conservation) Act 1980 was adopted to protect and conserve forests. The Act restricts the powers of the state with respect to the de-reservation of forests and the use of forestland for non-forest purposes (the term "non-forest purpose" includes clearing any forestland for the cultivation of cash crops, plantation crops, horticulture, or any purpose other than re-afforestation (The Forest (Conservation) Act, GOI, 1980).

8.4.4 The Air Act 1981

Ambient air quality standards were established in India under the Air Act 1981. This Act provides means for the control and abatement of air pollution. This Act seeks to reduce air pollution by avoiding the use of polluting substances and regulating appliances that cause air pollution. Under this Act, industrial consents are required from state boards for establishing and operating industries. It is the pollution control board's responsibility to implement the needful measures to reduce air pollution. This involves testing of the air in the pollution control areas, inspection of the pollution control equipment, and manufacturing processes (The Air Act, GOI, 1981).

In April 1994, CPCB notified the National Ambient Air Quality Standards (NAAQS) for major pollutants. These standards are designed with adequate safety margins to protect public health, vegetation, and property. The NAAQS prescribe specific standards for different sectors such as industrial, residential, rural, and others.

Industry-specific emission standards have also been developed for different industry segments such as iron and steel, cement, fertilizer, refineries, aluminum, etc. (The Air act, GOI, 1981).

The Air (Prevention and Control of Pollution) Amendment Act, 1987, was enacted to empower the central and state pollution boards to meet grave emergencies. With this act, the boards are authorized to take immediate measures to tackle emergencies and recover the expenses incurred from the offenders. The Act also empowers the boards to cancel consent for non-fulfillment of the conditions prescribed.

8.4.5   Environment (Protection) Act, 1986 (EPA)

This Act is an umbrella legislation designed to provide a framework for the co-ordination of central and state authorities established under the Water (Prevention and Control) Act, 1974 and Air (Prevention and Control) Act, 1981. Under this Act, the central government is empowered to take measures necessary to protect and improve the quality of the environment by setting standards for emissions and discharges; regulating the location of industries; management of hazardous wastes, and protection of public health and welfare (The Environment Protection Act, GOI, 1986).

From time to time, the central government issues notifications under the EPA for the protection of ecologically sensitive areas or issues guidelines for matters under the EPA.

8.4.6   Motor Vehicle Act 1988

The different objectives for which the motor Vehicles Act 1988 has been prepared are the following (Motor Vehicles Act, GOI 1988):

i.   To deal with the increasing number of commercial and personal vehicles in the country.
ii.   To facilitate the adoption of higher technology in the automotive sector.
iii.   To deal with a larger flow of passengers and freight so that islands of isolation are not created leading to regional or local imbalances.
iv.   To mitigate concerns related to road safety, pollution and to set control measures standards for transportation of hazardous and explosive materials.
v.   To simplify procedures and policies for private sector operations in the road transport field.
vi.   To provide effective ways of tracking down traffic offenders.
vii.   Rationalization of certain definitions particularly with respect to new types of vehicles.

viii. To monitor and control procedures relating to grant of driving licenses and its validity.

ix. Laying down of standards for the components and parts of motor vehicles.

x. Standards for anti-pollution control devices.

xi. Provision for issuing fitness certificates of vehicles.

xii. Enabling provisions for updating the system of registration marks.

8.4.7 Public Liability Insurance Act (PLIA), 1991

The Act covers accidents involving hazardous substances and insurance coverage for these. Where death or injury results from an accident, this Act makes the owner liable to provide relief as is specified in the Schedule of the Act. The PLIA was amended in 1992, and the Central Government was authorized to establish the Environmental Relief Fund, for making relief Payments (Public Liability Insurance Act (PLIA), GOI, 1991).

8.4.8 National Environment Tribunal Act, 1995

The National Environmental Tribunal Act, 1995, provided strict liability for damages arising out of any accident occurring while handling any hazardous substance and for the establishment of a National Environment Tribunal for effective and expeditious disposal of cases arising from such accident, with a view to give relief and compensation for damages to persons, property, and the environment, and for the matters connected therewith or incidental thereto (National Environment Tribunal Act, GOI, 1995).

8.4.9 The National Environmental Appellate Authority Act, 1997

This National Environmental Appellate Authority Act, 1997, provided for the establishment of a National Environment Appellate Authority. This authority is created to hear appeals with respect to restriction of areas in which any industry operation or process or class of industries are allowed to establish and operate to ensure safeguards under the Environment Protection Act 1986 (The National Environmental Appellate Authority Act, GOI, 1997).

8.5 Important Rules Related to Environment and Co-processing in India

To implement the provisions of the Acts to fulfil the desired objectives, the environment ministry at the central government and state governments have formulated and notified different rules with respect to each of the acts, and the same remain applicable in that state/central jurisdiction and need to be complied to.

Following are the MoEFCC notified rules that are specifically relevant in respect of the co-processing initiative:

a.   Hazardous and Other Waste Management Rules, 2016.
b.   Plastic Waste Management Rules, 2016.
c.   Solid Waste Management Rules, 2016.

The specific features pertaining to the co-processing initiative included in the above rules are elaborated below.

8.5.1 Hazardous and Other Waste (Management and Trans-Boundary Movement) Rules, 2016

The Hazardous Waste Management Rules were notified first time by the Environment Ministry of Government of India in 1989 (HWM Rules 1989). In these HWM Rules (1989), the major focus was scientific disposal of Hazardous waste through incineration and scientific landfilling. Facilities for incineration and landfilling got implemented in some of the states but not in all due to the Not In My Back Yard (NIMBY) syndrome and the objective of the rules was not getting fulfilled.

Subsequently, these rules were newly notified in 2008 with an inclusion of the provision of Rule 11 to permit the utilization of Hazardous waste if the same was scientifically and environmentally feasible to be implemented.

The HWM Rules were substantially modified and notified in 2016 with a focus towards sustainable management of wastes. A new category of wastes named as "Other" wastes was created. These included waste tyre, paper waste, metal scrap, used electronic items, etc. These wastes are recognized as a resource for recycling and reuse in these rules. They are termed as "Hazardous and Other Waste (Management and Trans-Boundary Movement) Rules, 2016. (HOWM Rules, GOI, 2016).

These rules included the following specific features:

• "Other Wastes" have been included.
• Waste Management as per waste management hierarchy has been proposed.
• Significantly revised forms for permitting import/export, filing of annual returns, undertaking transportation, etc., have been included.
• Standard Operating Procedures (SOPs) for safeguarding the health and environment from waste processing industry has been prescribed.

- Single window clearance for setting up of hazardous waste disposal facility and import of other wastes has been incorporated.
- Co-processing has been prescribed as a preferential mechanism for the management of wastes over disposal options.
- The approval process for co-processing of hazardous waste has been streamlined based on emission norms rather than trials.
- The process of import/export of waste under the Rules has been streamlined.
- Exemption methodology has been proposed for the import of metal scrap, paper waste, and various categories of electrical and electronic equipment for reuse.
- State Government responsibilities for environmentally sound management of hazardous and other wastes have been introduced.
- List of processes generating hazardous wastes has been aligned to the technological evolution in the industries.
- List of Waste Constituents with concentration Limits has been revised as per international standard and drinking water standard.
- The list of items prohibited for import has been included.
- State Pollution Control Board (SPCB) is mandated to prepare an annual inventory of the waste generated; waste recycled, recovered, utilized including co-processed.

8.5.2 Plastic Waste Management Rules 2016

Plastic Waste has been a cause of concern because of its adverse environmental impact. The plastic waste—that is not economically attractive or technically not feasible for recycling—tends to get littered in the environment. The plastic waste that is recyclable has economic value and hence has an appropriately operating value chain for recycling. The littered plastic waste not being a biodegradable material, tends to contaminate soil, water, and air causing survival concern to human beings, animals, and aquatic life due to its ingestion in the body through the food chain. Hence, this non-recyclable littered plastic waste needs to be appropriately managed. Most of this plastic waste is either packaging material or it is single use plastics.

It is estimated that around 25,940 tonne per day of plastic waste is generated in the country.

The government of India notified the first plastic waste management Rules in 1999. Subsequently, it notified the Plastic waste management rules in 2011 and then in 2016.

The Plastic Waste Management Rules, 2016, mainly addressed the non-recyclable packaging waste and have following specific features (PWM Rules, GOI, 2016):

- Increase the minimum thickness of plastic carry bags from 40 to 50 microns.
- Expand the jurisdiction of applicability from the municipal area to rural areas.
- Implement Extended Producer Responsibility (EPR) to plastic producers, importers, and brands.
- Introduction of Plastic waste management fee from producers, importers of plastic carry bags/multi-layered packaging, etc.

- To promote the use of plastic waste in different applications such as road construction, energy recovery, waste to oil, etc.

To address the concerns of the Single Use Plastics (SUPs), several state governments have also formulated their own rules and notifications and have notified them.

8.5.3   Solid Waste Management Rules, 2016

To address the concern of the Municipal Solid Waste generated in the country and its impact on the environment, Environment Ministry had formulated the Solid Waste Management Rules first time in the year 2000. The major focus of these rules was to dispose of the waste through composting of the biodegradable waste, Recycling of the recyclable waste and scientific landfills and incineration (with or without energy recovery) options for other waste.

Since the Solid Waste contains substantial quantity of resource value and recovery of the same is important from the sustainability point of view, Environment Ministry notified the new SWM Rules in 2016. (SWM Rules, GOI, 2016).

The salient features of SWM Rules 2016 are as follows:

- The jurisdiction has been increased beyond Municipal areas.
- The source segregation of waste has been mandated.
- Responsibilities have been introduced to segregate waste into Wet, Dry, and domestic hazardous wastes.
- Integration of waste pickers/rag pickers and waste dealers/Kabadiwalas in the formal system.
- Concept of "User Fee" and "spot fine" has been incorporated.
- Procedure to deal with used sanitary waste like diapers and sanitary pads has been incorporated.
- The concept of partnership in Swachh Bharat Abhiyan has been introduced.
- Responsibility has been fixed on all hotels and restaurants to segregate and treat the biodegradable waste.
- Responsibilities have been set for all resident welfare and market associations, gated communities, and institutions with an area > 5,000 sq. m. on managing the waste.
- New townships and Group Housing Societies have been made responsible to develop in-house waste handling and processing arrangements for biodegradable waste.
- Responsibilities have been set for every street vendor, developers of Special Economic Zone, industrial estate, industrial park, manufacturers of disposable products such as tin, glass, plastics packaging, etc.
- Responsibilities of all industrial units using fuel and located within 100 km from a solid waste-based RDF plant shall make arrangements to replace at least 5% of their fuel requirement by RDF.

- Non-recyclable waste having a calorific value of 1500 kcal/kg or more shall not be disposed of on landfills.
- High calorific wastes shall be used for co-processing in cement or thermal power plants.
- Construction and demolition waste should be stored, separately disposed of, as per the C&D Waste Management Rules, 2016.

As of 2018, 55,913 wards (out of the total 82,842 wards) are covered by 100% door-to-door collection. 22.85% of the total waste generated is currently being processed. Currently, there are 9 functional waste-to-energy (WTE) plants and 148 waste-to-compost (WTC) plants that are operational across the country (MoHUA, 2018) new_AR-2017–18 (Eng)-Website.pdf (mohua.gov.in).

8.6   Environment Policy Framework of India and Option of Co-processing

The present national policies in India for environmental management are contained in the National Forest Policy, 1988, National Conservation Strategy and Policy Statement on Environment and Development, 1992, Policy Statement on Abatement of Pollution,1992. Some sector policies such as the National Agriculture Policy, 2000 National Population Policy, 2000 National Water policy 2002 have also contributed towards environmental management. All these policies have recognized the need for sustainable development in their specific contexts and formulated necessary strategies to give effect to such recognition. The National Environment Policy seeks to extend the coverage and fill in gaps that still exist, considering present knowledge and accumulated experience. It does not displace but builds on the earlier policies. To achieve growth in a sustainable manner, the government of India introduced several policy measures in 2016.

Co-processing in cement kilns was one of the important options that was introduced in the policy framework of India notified in the year 2016 for the management of non-recyclable wastes derived out of Industrial and Municipal activities and non-cattle feed biomass generated while undertaking agricultural activities.

The options for waste management specified in the rules notified prior to 2016 were based on the principle of disposal. Landfill and incineration were the proposed options in those rules. The new rules notified in 2016 are based on the principle of sustainability. As per earlier rules, these materials used to get dumped or landfilled and their resource value was getting wasted. Co-processing of these waste streams helps the substitution of fossil fuels and natural raw materials in the cement plants reducing the GHG emissions and also conserving the natural materials.

8.7   Co-processing Related Documents Published by Different Agencies Internationally

Following documents have been published at the international level to promote co-processing:

A.   Guidelines on co-processing Waste Materials in Cement Production published through GTZ Holcim Public–Private Partnership in 2006 (GTZ/Holcim, 2006).
B.   Technical guidelines on the environmentally sound co-processing of hazardous wastes in cement kilns published by Basel Convention in 2011 (UNEP, 2011).
C.   Guidelines for Co-Processing Fuels and Raw Materials in Cement Manufacturing published by WBCSD in 2014 (WBCSD, 2014).
D.   Guidelines on pre-processing and co-processing of wastes in cement production published by GIZ/LafargeHolcim in 2020 (GIZ/LafargeHolcim, 2020).

8.8   Recognition of Co-processing by Different Global Bodies

Following different global bodies have provided recognition to co-processing.

8.8.1   UNEP

United Nations Environment Program (UNEP), through the Basel Convention, has developed technical guidelines on the environmentally sound co-processing of hazardous waste in cement kilns.

8.8.2   SINTEF

Various researchers at SINTEF, the largest independent research organization in Scandinavia, consider co-processing as a problem solver and support the use of local cement kilns for the environmentally sound destruction of hazardous organic chemicals like toxic pesticides and persistent organic pollutants in several countries.

8.8.3 GIZ

GIZ operates worldwide and assists the German government in achieving its objectives in the field of international cooperation, especially regarding sustainable development and resource management. In 2018, the organization together with Lafarge-Holcim, updated its guidelines for an efficient and environmentally sound preprocessing (preparation of the waste to make it suitable for its treatment in cement kilns) and co-processing (recycling and recovery process) activities.

8.8.4 ADEME

The French Environment and Energy Management Agency has approved the technical reliability of co-processing. ADEME promotes co-processing as a waste treatment solution for public bodies in charge of waste management.

8.8.5 MOEFCC, Government of India

In 2016, the Indian Ministry of Environment, Forest and Climate Change, and Central Pollution Control Board recognized that "there is dual benefit in co-processing waste in cement kilns, in terms of utilizing the waste as a supplementary fuel as well as an alternative raw material" and stated that "co-processing in cement kilns is considered an environmentally friendly option for managing different kinds of waste".

8.8.6 Five-Year Plan on Ecology and Environment Protection, China Government

In the context of setting up the 13th Five-Year Plan on Ecology and Environment Protection (November 2016), the Chinese government encourages the development of waste treatment infrastructures. Within this framework, the co-processing of municipal solid waste and hazardous waste is promoted. By 2020, the aim is to significantly increase the number of cement plants equipped to co-process waste.

8.9   Co-processing Related Regulations in Different Countries

Co-processing of wastes in cement kilns is being practiced for more than three decades. In the earlier days, it used to be practiced in a few countries. Currently, it is practiced in many different countries. When appropriate regulations are not in place, co-processing can cause significant negative impacts on human health and environment. Effective regulatory frameworks are essential to have beneficial co-processing practices. Several countries have established a regulatory framework for co-processing and have refined them several times through amendments subsequently. Following are the regulatory frameworks of some of the European countries, as well as other countries, such as Japan, United States, Australia, Brazil, and South Africa. These different countries have created their own regulatory framework to permit co-processing in the cement plant located in their countries. These regulations are briefly discussed in this section (Hasanbeigi et al., 2012).

Many countries around the world have established emission limits for different types of pollutants from co-processing plants, some of which are described below. The EU WID establishes limits on the emissions of heavy metals, dioxins and furans, CO, dust, total organic carbon, HCl, HF, SO2, and NOx from co-processing plants. Dioxins and furans must be measured at least twice per year, and at least every 3 months for the first 12 months of a plant's operation (Karstensen, 2008).

Dust from de-dusting equipment can be partially or totally recycled into cement manufacturing processes. If recycling is not feasible or not allowed, the dust must be evaluated before use in soil or waste stabilization or for agricultural purposes (GIZ/Holcim, 2006). If dust is landfilled, the landfill design must use BAT. In most EU countries, test burns are usually conducted to evaluate the performance of new technology or process to reduce emissions; the quality of the resulting clinker is also evaluated to ensure that hazardous residues from the waste-burning process do not leach from the final product and pose an environmental hazard (GIZ/Holcim, 2006).

8.9.1   European Union

Through the Waste Framework Directive (2008/98/EC) (WFD), The European Union has defined its basic waste policy. All member states of the European Union are desired to align their national policies with the WFD directive within a defined period of time. Basic concepts and definitions, including waste avoidance, reuse, recycling, recovery, and management are the main inclusions in the WFD. The WFD also defines the waste management hierarchy that prioritizes sustainable approaches such as avoidance of waste generation as against the unsustainable ones such as land filling. Cement kiln co-processing is regarded in WFD as resource recovery and is thus prioritized over incineration or landfilling. Therefore, waste avoidance, reuse, recycling, and recovery of wastes do not compete with co-processing. The WFD has also established the "extended producer responsibility" and "polluter pays" principles. This provides a considerable incentive for the cement industry to undertake co-processing. Due to these principles, waste-producing and waste-handling agencies need to pay the cement industry the co-processing services extended by it (EU, 2008).

One of the important drivers for the cement industry to undertake co-processing in Europe was the Landfill Directive (1999/31/EC), which was established in 1999. As per this directive, the members of the European Union are required to establish national strategies to reduce the landfilling requirements. As per this directive, Sweden banned landfilling of separated combustible wastes in 2002 and organic wastes in 2005. Because of this directive, for the wastes that cannot be reused or recycled, alternative options of incineration and co-processing have started getting implemented. European Commission have also formulated the Waste Incineration Directive (WID) (2000/76/ EC) in 2000, which addresses the public concerns related to the environmental and health impacts of burning waste. The WID has laid out methodologies for granting permits for delivery and reception of waste, water discharges, operational conditions, air emissions limits, residues, monitoring and surveillance, access to information and public participation, reporting, and penalties, etc., for incineration and co-processing. The WID directive imposes stringent regulations on emissions, operational conditions, and technical requirements than those that were previously in place.

The Integrated Pollution Prevention and Control (IPPC) Directive has been framed by European Union which also applies to co-processing plants. This directive provides an integrated approach for the application of "best available techniques" (BATs) and environmental permitting system (EIPPCB, 2006).

In most EU member states, regulatory and enforcement responsibility is divided among several different "competent authorities." In several member states, such as Austria, Germany, Belgium, and Bulgaria, regulatory functions are divided between the national/federal level and the regional/state level. In other countries, such as in Denmark and Hungary, regional authorities carry out the major control functions for industrial installations. Regulatory functions are carried out at the municipal/local authority level in the Czech Republic, Netherlands, UK, and Ireland (Milieu, 2011).

8.9.2 Japan

Japan's Waste Management and Public Cleaning Law was established in 1970. During the past decade, Japan has developed an integrated waste and material management approach that promotes dematerialization and resource efficiency. Landfill shortage and dependency on imported natural resources have been key drivers of these changes. The 2000 Basic Law for Establishing a Sound Material-Cycle Society. Also, on December 7, 2009, EPA signed two distinct findings (Endangerment Finding and Cause or Contribute Finding) regarding greenhouse gases under Sect. 202(a) of the Clean Air Act. Endangerment Finding indicates that six GHGs threaten the public health and welfare of the current and future generation. However, these findings do not themselves impose any requirements on industry or other entities that have integrated the environmentally sound management of waste with the "3R" (reduce, reuse, and recycle) approach. This represents a shift in emphasis from waste management to sound materials management.

The eco-towns policy is how cement co-processing has been directly incorporated into industrial planning policies in Japan in recent years. The eco-town concept originated through a subsidy system established by the Japanese Ministry of Economy, Trade, and Industry and the Ministry of the Environment in 1997.

8.9.3   United States of America

In the U.S., MACT standards are established under Sect. 112 of the Clean Air Act through the national emissions standards for HAPs. The MACT standards, such as the Portland Cement Kiln MACT, are intended to achieve "the maximum degree of reduction in emissions," while taking into account cost, non-air-quality health and environmental impacts, and energy requirements. Emissions standards for the U.S. cement industry are specified in the Code of Federal Regulations 40, Part 60, Subpart F. The standards apply to kilns, clinker coolers, raw mill systems, finish mill systems, raw mill dryers, raw material storages, clinker storages, finished product storages, conveyor transfer points, bagging, and bulk loading and unloading systems Under the authority of Sect. 129 of the Clean Air Act, the U.S. EPA has proposed rulemaking for commercial and industrial solid waste incineration units (CISWI), which potentially include co-processing cement plants. The CISWI MACT standards were released and now the U.S. EPA is in the process of reconsideration based on feedback.

The U.S. regulations for co-processing were largely the result of concerns related to environmental protection and the implementation of the 1970 Clean Air Act. In the United States, the U.S. EPA regulates emissions from the U.S. cement industry co-processing or delegates this authority to state or local agencies. In 2008–2010, the U.S. EPA established the national "New Source Review/Prevention of Significant Deterioration (NSR/PSD)" enforcement initiative for the cement industry. The initiative was continued in the form of the national initiative "Reducing Air Pollution from the Largest Sources" for the years 2011–2013.

8.9.4   Brazil

National Regulatory Act No. 264/99 of Brazil establishes technical and operational criteria, emissions limits, and pre-permit testing requirements for co-processing permits for cement kilns. National Regulatory Act 316/02 (Licensing of Incineration/Co-incineration), establishes limits for emissions of dioxins and furans

(0.5 Nanograms per cubic nanometre [ng/Nm3] from cement kiln co-processing (Maringolo, 2007).

8.9.5 South Africa

National Policy on the Thermal Treatment of General and Hazardous Waste (the South Africa National Policy) relies on the EU Incineration Directive 2000/76/EC (especially for air emissions limits) and other international policies, including co-processing guidelines by WBCSD and Holcim, as models.

8.9.6 China

The various Chinese policies that promote the cause of co-processing are the following:

- The Development Policy of Cement Industry (NDRC, 2006).
- Technical Policy on Pollution Prevention and Control for Co-processing Solid Waste in Cement Kilns, 2016.
- Industry Green Development Plan (2016–2020).
- Planning for the Innovative Capacity Development of Industrial Technology (2016–2020).

Based on the above defined policy support, by the end of 2018, there had been 57 CKC production lines for MSW treatment distributed in 16 provinces. The top 3 provinces with the highest treatment capacity are Guizhou (2613 t/d), Guangxi (2500 t/d), and Anhui (1900 t/d). Green Manufacturing Specific Action Plan, 2016— The launch of pilot projects of co-processing MSW in cement kilns was one of the actions in 2016's Green Manufacturing action plan. Circular economy development strategies and near-term actions encourage cement kilns to co-process solid wastes for resource utilization (Kosajan et al., 2021).

Industry green development plan (2016–2020) encourages the implementation of co-processing of solid waste in cement kilns according to local conditions. Planning for the Innovative Capacity Development of Industrial Technology (2016–2020) supports the research, development, and implementation of the complete sets of technologies and equipment for co-processing waste in cement kilns. Technical policy on pollution prevention and control for co-processing solid waste in cement kilns. Establish the technical requirements for different implementation scenarios of CKC. Notice on accelerating industrial energy conservation and green development. Develop and improve the green finance instruments to further support the projects for comprehensive utilization of resources, e.g., co-processing solid waste in cement kilns (Hasanbeigi et al., 2012).

8.10   Conclusion

For implementing pre-processing of wastes into AFRs and co-processing them in cement kilns in a trouble free, environmentally sound, and ecologically sustaining manner, appropriate regulations and standards are required. These are needed in five key areas: environmental performance, product quality, waste quality, operational practices, and safety and health requirements for employees and local residents. Many countries have drafted their regulatory framework by addressing the important aspects of these five years. This chapter has reviewed briefly the various regulatory provisions made by different countries that are practicing co-processing and the Indian regulatory framework in detail.

References

EIPPCB (European Integrated Pollution Prevention and Control Bureau). (2006). Reference Document on Best Available Techniques for the Waste Treatment Industries. European Commission. http://eippcb.jrc.es/reference/wt.html

Environment (Protection) Act (EPA). (1986). https://legislative.gov.in/sites/default/files/A1986-29.pdf

EU. (2008). DIRECTIVE 2008/98/EC of the European Parliament and of the Council of 19 November 2008 on waste and repealing certain Directives, Official Journal of the European Union.

GIZ/LafargeHolcim. (2020). Guidelines on pre-processing and co-processing of wastes in cement production. https://www.giz.de/de/downloads/giz-2020_en_guidelines-pre-coprocessing.pdf

Ghosh, S.K. (2020). Circular Economy in India, book, Circular Economy: Global Perspective. Springer Nature. https://doi.org/10.1007/978-981-15-1052-6

GTZ/Holcim. (2006). Guidelines on co-processing of wastes in cement production

Hasanbeigi, A., Lu, H., Williams, C., & Price, L. (2012). International best practices for pre-processing and co-processing municipal solid waste and sewage sludge in the cement industry.

HOWM Rules. (2016). Notified by MoEFCC Government of India (GSR 395 E) dated 4 April 2016. http://moef.gov.in/wp-content/uploads/2017/08/GSR-395E.pdf

Karstensen, K. H. (2008). Formation, release and control of dioxins in cement kilns-A review. *Chemosphere, 70*(2008), 543–560.

Kosajan, V., Wen, Z., Zheng, K., Fei, F., Wang, Z., & Tian, H. (2021). Municipal solid waste (MSW) co-processing in cement kiln to relieve China's MSW treatment capacity pressure. https://doi.org/10.1016/j.resconrec.2020.105384

Milieu Law & Policy Consulting Ltd (Milieu). (2011). Provisions on penalties related to legislation on industrial installations: Document on good practices. Brussels, Belgium.

Maringolo, V. (2007). Use of waste-derived fuels and materials in the Brazilian cement industry. *Global Fuels Magazine*:33–34.

MoHUA, Annual Report 2017–2018. http://mohua.gov.in/upload/uploadfiles/files/new_AR-2017-18%20(Eng)-Website.pdf

Motor Vehicles Act. (1988). https://www.morth.nic.in/motor-vehicles-act-1988-0

PWM Rules. (2016). Notified by MoEFCC, Government of India (GSR 320 E) dated 18 March 2016 PWM-Rules-2016-English.pdf (moef.gov.in).

Public Liability Insurance Act (PLIA). (1991). GOI. https://legislative.gov.in/sites/default/files/A1991-06.pdf

SWM Rules. (2016). Notified by Mo"EFCC, Government of India (SO1357 E) dated 8 April 2016. SWM-2016-English.pdf (moef.gov.in).

The Air act. (1981). https://legislative.gov.in/sites/default/files/A1981-14.pdf

The Forest (Conservation) Act. (1980). https://legislative.gov.in/sites/default/files/A1980-69_0.pdf

The National Environmental Appellate Authority Act. (1997). GOI. https://legislative.gov.in/sites/default/files/A1997-22_0.pdf

The factories Act. (1948). GOI. https://www.indiacode.nic.in/bitstream/123456789/1530/1/AAA 1948_63.pdf

The Water Act. (1974). https://legislative.gov.in/sites/default/files/A1974-6.pdf

UNEP. (2011). Technical Guidelines on environmentally sound co-processing of Hazardous wastes in cement kilns

WBCSD. (2014). Guidelines for Co-Processing Fuels and Raw Materials in Cement Manufacturing. CSI_Co-Processing_Fuels_and_Raw_Materials.pdf (wbcsd.org)

Part V
Co-processing, Pre-processing and AFR in Cement Kiln: Operations, Maintenance and Emission Controls

Chapter 9
Emission Considerations in Cement Kiln Co-processing

9.1 Introduction

Thermal treatment of every combustible or volatile material causes emissions. The nature and characteristics of these emissions depend upon the chemical nature of the material being combusted or volatilized. Emissions can be of different types and their impact on the environment or living beings can be varying and disastrous sometimes.

It is important therefore that the thermal treatment needs to be implemented in such a manner that emissions that are harmful to the living beings or environment are avoided or maximally reduced. Therefore, different countries have enacted emission standards to ensure that the thermal treatments are implemented with the desired level of caution to reduce impacts on health and environment. Countries revise these emission standards progressively in a downward direction for improved environmental protection.

Co-processing is a thermal treatment process, and therefore, there is an impact of the same on the emissions from the cement manufacturing process. There is a need to understand these aspects well to ensure that they are monitored and controlled appropriately to comply with specified standards.

9.2 Different Kinds of Emissions Encountered During Thermal Treatment of Fossil Fuels and Wastes

There are different types of emissions that take place during the thermal treatment of materials, such as, particulate emissions, acidic emissions, GHG emissions, toxic emissions, poisonous emissions, VOC emissions, Heavy metal emissions, Dioxin and Furan emissions, etc. These gases are released from the thermal processes while using both fossil fuel/resources and alternative fuel/resources.

© The Author(s), under exclusive license to Springer Nature Singapore Pte Ltd. 2022 179
S. K. Ghosh et al., *Sustainable Management of Wastes Through Co-processing*,
https://doi.org/10.1007/978-981-16-6073-3_9

9.2.1 Particulate Emissions

The particulate emissions or dust emissions from cement plants are of two kinds—Stack and fugitive emissions. The stack emissions occur due to the inefficiencies associated with the dust control devices such as Cyclones, Electro Static Precipitators (ESP)s, Baghouses installed on the raw mills, cement kilns, cooler, grinding mills, and packing plants. The dust emission from all the stacks needs to be appropriately monitored and controlled.

The fugitive emissions occur on account of inefficiencies associated with the dust control devices such as dust extraction and dust suppression systems installed at the storage and material handling and transfer stations. In the case of co-processing operation, the particulate emissions of relevance are the kiln stack emissions and these are measured as $PM_{0.1}$, PM_1, $PM_{2.5}$, and PM_{10}. The major concentration of the particulate emissions generally includes $PM_{2.5}$ and PM_{10} fractions. The particulate emissions in the kiln stack consist of raw material constituents and also various heavy metals (Gupta et al., 2012).

Usually, kiln dust is completely returned to the process—either to the kiln system or to the cement mill. Kiln dust is highly alkaline and may contain trace elements such as heavy metals corresponding to the contents in the source materials. Bypass dust extracted from the kiln system may be highly enriched in alkalis, sulphates, and chlorides and—similarly to filter dust—in some cases cannot be completely recycled to the process.

Impact

The particulate emissions affect the life and wellbeing of workers, children, and people in close communities as well as the flora and fauna. Diseases such as chronic obstructive pulmonary, silicosis, preterm delivery, psychasthenia, endocrine disruption, cancer, infertility, etc., are associated with these pollutants (Adeyanju and Okeke 2019). Cement industry-derived pollutants appear to play multiple roles in stimulating abiotic stress responses in plants. Cement dust deposition on agriculture fields can affect soils, photosynthesis, transpiration, and the respiration of plants (Shah et al., 2020). Also, Cement dust depositing on the nearby water bodies impacts the quality of the water and the marine species.

Monitoring and Control

Considering the impact of particulate emissions on the health and environment, it is important to control the level of these emissions from the cement kiln stacks. For achieving this control, five different types of devices are utilized, namely Gravity Settling Chambers, Mechanical collectors, Particulate Wet Scrubbers, Electro Static Precipitators, and Fabric Filters (Zulfiqar et al., 2013).

9.2.2 Acidic Emissions

These emissions are those that tend to get converted into acids while combining with water vapour from the environment and tend to cause acid rain. These consist of SO_2, SO_3, HCl, HF, NO, NO_2, P_2O_5, H_2S, CO_2, etc. These are released in the thermal process when the relevant chemical constituent is present in the fossil or alternative material. For example, the presence of Sulphur releases H_2S, SO_2, and SO_3; the presence of Chlorine causes HCl emissions; the presence of Fluorine causes HF emissions, the presence of Nitrogen causes NOx emissions, presence of the Carbon causes CO_2 emissions.

NOx formation is an inevitable consequence of the high-temperature combustion process, with a smaller contribution resulting from the chemical composition of the fuels and raw materials. Nitrogen oxides are formed by oxidation of molecular nitrogen in the combustion air ("thermal" NOx is the sum of nitrogen oxides; in cement kiln exhaust gases, NO and NO_2 are dominant, >90% NO and <10% NO_2). Thermal NOx formation is dependent on the combustion temperature with a marked increase above 1400 °C. In the secondary firing of a preheater/precalciner kiln with a flame temperature of not more than 1200 °C, the formation of thermal NOx is much lower compared to the main burner flame. Up to 60% of the total fuel can be burnt in the calciner. NOx emissions in cement kilns (expressed as NO_2) typically vary between 300 and 2000 mg/m^3.

Sulphur compounds enter the kiln system either with the fuels or with the raw materials. Sulphides and organic sulphur compounds in raw materials are pyritic in nature. It gets decomposed and oxidized at moderate temperatures of 400 to 600 °C to produce SO_2 when the raw materials are heated by the exhaust gases. At these temperatures, not enough calcium oxide is available to react with the SO_2. Therefore, in a dry preheater kiln, about 30% of the total sulphide input may leave the preheater section as gaseous SO_2. The emission limits from the cement plants, therefore, are linked to the pyritic sulphur content in the raw materials. The allowance limits for SOx emissions for cement plants have been proposed by MoEFCC—Government of India—as 100, 700, and 1000 mg/Nm^3, for the pyritic sulphur content in the limestone of < 0.25%, 0.25 to 0.5%, and more than 0.5%, respectively. During direct operation, i.e., with the raw mill off—most of it is emitted to the atmosphere. During compound operation, i.e., with the raw mill online—typically between 30 and 90% of that remaining SO_2 is additionally adsorbed to the freshly ground raw meal particles in the raw mill ("Physico-Chemical absorption"). Sulphur compounds in raw materials, namely sulphates, are thermally stable up to temperatures of 1200 °C, and will thus enter the sintering zone of the rotary kiln where they are decomposed to produce SO_2. Part of the SO_2 combines with alkalis and is incorporated into the clinker structure. The remaining part of SO_2 is carried back to the cooler zones of the kiln system where it reacts either with calcined calcium oxide or with calcium carbonate, thus being reintroduced to the sintering zone again, called chemical SO2 absorption.

Inorganic and organic Sulphur compounds introduced with the fuels will be subject to the same internal cycle consisting of thermal decomposition, oxidation to SO_2, and reaction with alkalis or with calcium oxide. With this closed internal cycle, all the Sulphur which is introduced via fuels or raw material sulphates will leave the kiln chemically incorporated in clinker and will not give rise to gaseous SO_2 emissions.

Impact

Acid gases cause significant damage to the pipelines or stacks through which they run. They bring about extensive corrosion and rusting, while they can also significantly shorten the lifespan of both pipelines and stacks replacement. The repair of these components is significantly cost intensive. In addition to their corrosive nature, acid gases can also cause serious health complications among humans. Prolonged exposure to acid gases can bring about severe illness or exacerbate existing conditions, even causing death in the most extreme situations. Finally, the release of acid gases into the atmosphere can do untold damage to the environment, manifesting itself in such phenomena as acid rain and global warming.

Monitoring and Control

To monitor these emissions during co-processing, it is important to have the necessary and appropriate stack monitoring system installed on the stacks of the cement kilns. There are different online stack monitoring devices available using which it is possible to measure these various Acidic emissions online.

In the co-processing operation, the acidic gases that get generated are SOx, NOx, HCl, HF, CO_2, and P_2O_5. Some of the acidic gases generated during thermal treatment of the fossil as well as alternative fuels such as HCl, HF, P_2O_5, SO_2, and SO_3 get fully picked up due to their reaction with the limestone present in the raw meal being fed into the cement kiln. They get fully absorbed and reacted with the finely ground limestone powder. Hence, these acid emissions are not a cause of concern in cement kiln.

Only acidic gases that do not get reacted with the limestone are the NOx and CO_2 emissions. Control of CO2 is explained in the next section dealing with GHG emissions. The NOx emissions need to be controlled through another technology which is called DeNOx.

There are two technologies for reducing NOx using DeNOX technology. The first one is Selective Catalytic Reduction (SCR) Technology. In this technology, NH_3 gas is used to react with NOx over a bed of catalyst at a temperature of 300–400 degree C to convert it into Nitrogen. This technology is utilized substantially in Thermal Power Plants and is not so much used in the cement industry. The other technology is Selective Non-Catalytic Reduction (SNCR) technology. In this technology, Aqueous Ammonium Hydroxide (NH_4OH) / Urea solution (CH_4N_2O) is used to react with NOx at about 1000 degree C to convert it into Nitrogen.

9.2.3 GHG Emissions

These emissions are those that cause the climate change impact. These include CO_2, CH_4, HCl, N_2O, CFC, HFC, etc. In the thermal treatment of fossil and alternative materials, the GHG emission of major concern is CO_2.

Carbon dioxide emissions arise from the calcination of raw materials and the combustion of fossil fuels. CO_2 resulting from calcination can be influenced to a very limited extent only. Emissions of CO_2 resulting from fuel combustion have generally been reduced due to the strong economic incentive for the cement industry to minimize fuel energy consumption. Nearly 30% of CO_2 reduction takes place in the last 25 years arising mainly from the adoption of more fuel-efficient kiln processes leaving little scope for further improvement. Potential is mainly left to the increased utilization of renewable alternative fuels or other waste-derived fuels and the production of blended cement with mineral additions substituting clinker.

Impact

CO_2 in the GHG emission is the major cause of concern currently from the climate change point of view. At current emission rates, temperatures in the atmosphere could increase by 2 °C, which the United Nations Intergovernmental Panel on Climate Change (IPCC) designated as the upper limit to avoid "dangerous" levels, by 2036.

Monitoring and Control

CO_2 emissions from the Cement manufacturing process are calculated by using CSI-WBCSD protocol. The cement Industry pursues various levers like Clinker factor reduction, thermal efficiency improvement, enhanced utilization of alternative fuels and raw materials, enhanced utilization of renewable electrical energy, utilization of waste heat recovery systems, improvement of process efficiency, and adoption of the latest technologies from time to time to control/reduce CO_2 emissions from the process. However, to achieve the desired reduction in CO_2 emissions, cement industry also has to adopt the secondary abatement technologies such as Carbon Capture and Storage (CCS) that are at different stages of development.

9.2.4 VOC Emissions

Natural raw materials such as limestone's, marls, and shales may also contain up to 0.8 % w/w of organic matter—depending on the geological conditions of the deposit. A large part of this organic matter may be volatilized in the kiln system even at moderate temperatures between 400 and 600 ° degrees C. Kiln tests with raw meals prepared from raw materials of different origins have demonstrated that approximately 85% to 95% of the organic matter present in them is converted into CO_2 in presence of 3% excess oxygen and 5 to 15% is oxidized to CO. A small proportion, usually less than 1% of the total organic carbon ("TOC") content may be emitted as volatile organic compounds ("VOC") such as hydrocarbons.

The emission level of VOC in the stack gas of cement kilns is usually between 10 and 100 mg/Nm^3, with a few excessive cases up to 500 mg/Nm^3. The CO concentration in the clean gas can be as high as 1000 mg/Nm^3, even exceeding 2000 mg/Nm^3 in some cases. The carbon monoxide and hydrocarbon contents measured in the stack gas of cement kiln systems are essentially determined by the content of organic matter in the raw materials and are, therefore, not an indicator of incomplete combustion of conventional or alternative fuels. Organic matter introduced to the main burner and the secondary firing will be destroyed due to the high temperatures and the long retention time.

Impact

VOC coming from raw material sources such as limestone, etc., will be observed in the kiln stack. The same generally is not toxic but still needs to be controlled to the desired levels.

Monitoring and Control

The best approach is to reduce or eliminate the use of organics containing raw materials. However, in case the extent of VOC is very high, the feasibility of implementing technologies such as flaring, or adsorption may need to be explored for implementation.

9.2.5 Heavy Metal Emissions

The various heavy metals of concern in the emissions are Mercury (Hg), Cadmium (Cd), Thallium (Tl), Antimony (Sb), Arsenic (As), Lead (PB), Cobalt (Co), Chromium (Cr), Copper (Cu), Manganese (Mn), Nickel (Ni), Vanadium (V), Zinc (Zn), Selenium (Se), and Tin (Sn) and their compounds (CPCB, 2010).

Impact

Heavy metals can bind to vital cellular components such as structural proteins, enzymes, and nucleic acids, and interfere with their functioning. Symptoms and effects can vary according to the metal or metal compound, and the dose involved. Broadly, long-term exposure to toxic heavy metals can have carcinogenic, central and peripheral nervous system, and circulatory effects (Lanids et al. 2000).

Monitoring and Control

Monitoring of the heavy metals in the emissions is carried out by sampling the stack gases and determining their concentrations using various instruments such as Atomic Absorption, X-Ray fluorescence, Chemical analysis, etc. They need to be controlled by ensuring their concentrations in the input streams.

9.2.6 Dioxin and Furan Emissions

Persistent organic pollutants (POPs) can be unintentionally produced and emitted from waste co-processing in cement kilns. Detailed study of POP formation and emission by cement kilns co-processing is needed to assess to know the potential risks. Researches [(Yang et al., 2019, Shibamoto et al., 2007, Liu et al., 2015, 2016, Mukherjee et al., 2016)], studied in fields and laboratory simulation experiments to investigate the formation and release of polychlorinated dibenzo-p-dioxins and dibenzofurans (PCDD/Fs). However, the formations, characteristics, and emission factors of various emerging unintentionally produced POPs (PCDD/Fs and poly-brominated dibenzo-p-dioxins and dibenzofurans, polychlorinated naphthalenes, and chlorinated and brominated polycyclic aromatic hydrocarbons) in cement kilns co-processing have been comprehensively reviewed by a few researchers (Zou et al., 2018, Karstensen, 2008). Data from field studies indicated that the main stages in which POPs are unintentionally produced in cement kilns co-processing solid waste are the cyclone preheater outlet, suspension preheater boiler, humidifier tower, and back-end bag filter. The raw material composition, chlorine and bromine contents, and temperature are the most important factors affecting POP formation.

Persistent Organic Pollutants (POPs) are organic chemical substances, that is, they are carbon-based. They possess a particular combination of physical and chemical properties such that, once released into the environment, they:

- remain intact for exceptionally long periods of time (many years);
- become widely distributed throughout the environment as a result of natural; processes involving soil, water and, most notably, air;
- accumulate in the fatty tissue of living organisms including humans, and are found at higher concentrations at higher levels in the food chain; and
- are toxic to both humans and wildlife.

As a result of releases to the environment over the past several decades due especially to human activities, POPs are now widely distributed over large regions (including those where POPs have never been used) and, in some cases, they are found around the globe. This extensive contamination of environmental media and living organisms includes many foodstuffs and has resulted in the sustained exposure of many species, including humans, for periods of time that span generations, resulting in both acute and chronic toxic effects.

In addition, POPs concentrate on living organisms through another process called bioaccumulation. Though not soluble in water, POPs are readily absorbed in fatty tissue, where concentrations can become magnified by up to 70,000 times the background levels. Fish, predatory birds, mammals, and humans are high up the food chain and so absorb the greatest concentrations. When they travel, the POPs travel with them. As a result of these two processes, POPs can be found in people and animals living in regions such as the Arctic, thousands of kilometers from any major POPs source.

Specific effects of POPs can include cancer, allergies and hypersensitivity, damage to the central and peripheral nervous systems, reproductive disorders, and disruption of the immune system. Some POPs are also considered to be endocrine disrupters, which, by altering the hormonal system, can damage the reproductive and immune systems of exposed individuals as well as their offspring; they can also have developmental and carcinogenic effects.

PCDD/PCDFs are produced unintentionally due to incomplete combustion, as well as during the manufacture of pesticides and other chlorinated substances. There are 75 different dioxins (PCDD) and 135 different types of furans (PCDF).

Dioxins and Furans have been associated with a number of adverse effects in humans, including immune and enzyme disorders and chloracne, and they are classified as possible human carcinogens. Food (particularly from animals) is the major source of exposure for humans.

Impact

Dioxins are persistent and not easily degraded by environmental microbes. They tend to accumulate in the environment. Dioxins are substantially more soluble in lipids than in water. Their concentration keeps on magnifying at each trophic level. This leads to high concentrations at the highest trophic levels which are seals and predatory birds. The human diet is quite diverse and hence, its concentrations in humans are not as high as in the most endangered wild species. Dioxin-like compounds are regulated strictly since the 1980s because of their toxicity and persistence (Jouko, 2019).

Monitoring and Control

For new cement kilns and major upgrades, the BAT for the production of cement clinker is a dry process kiln with multi-stage preheating and precalcination. A smooth and stable kiln process, operating close to the process parameter set points is beneficial for all kiln emissions as well as the energy use. PCDD/PCDF control in cement production becomes a simultaneous effort to reduce the precursor/organic concentrations, preferably by finding a combination of optimum production rate and optimum gas temperatures and oxygen level at the raw material feed end of the kiln, and reducing the APCD temperature. Feeding of alternative raw materials as part of raw-material-mix should be avoided if it includes elevated concentrations of organics and no alternative fuels should be fed during start-up and shut down. The most important measure to avoid PCDD/PCDF formation in wet and long dry kilns seems to be quick cooling of the kiln exhaust gases to lower than 200 °C. Modern preheater and precalciner kilns have this feature already inherent in the process design and have APCD temperatures less than 150 °C. Operating practices such as minimizing the build-up of particulate matter on surfaces can assist in maintaining low PCDD/PCDF emissions.

9.2.7 Other Toxic Emissions

Combustion and thermal processes are dominant sources of toxic air pollution. They also produce chronically toxic products of incomplete combustion (PICs) such as benzene, polychlorinated dibenzo-*p*-dioxins and dibenzofurans (PCDD/Fs), acrylonitrile, and methyl bromide, PAH, VOCs, etc. They are also PAH and Although these toxic combustion by-products are formed in many types of combustion and thermal processes, they have historically been of particular concern for incineration of hazardous wastes and soils/sediments contaminated with hazardous wastes. Further, low- or moderate-temperature treatment has the potential to form more toxic by-products than incineration (Stephania et al., 2006).

Impact

These emissions mainly impact health. Diseases caused due to these substances include principally respiratory problems, such as Chronic Obstructive Pulmonary Disease (COPD), asthma, bronchiolitis, and also lung cancer, cardiovascular events, central nervous system dysfunctions, and cutaneous diseases, etc (Ioannis et al., 2020).

Monitoring and Control

The monitoring of these emissions is carried out using online emission monitoring systems and with improved combustion processes having a higher temperature, sufficient residence time and proper control over operating parameters. Sometimes post-combustion treatment processes may be required which include adsorption on activated carbon, etc.

9.3 Description of the Cement Kiln Co-processing Operation

The cement kiln system consists of a rotary kiln shown in brown colour which is slightly inclined and is rotating at a defined speed. Figure 9.1 illustrates the cement kiln co-processing operation. At one end, fossil fuel such as coal, oil or gas is fired with a regulated quantum of air. The temperature of the flame is generally maintained above 1800 °C and the residence time of the gases above 1100 °C is 6–8 seconds. The temperature at the calciner is above 850 °C.

The hot gases from the fuel burning process travel to the other side of the kiln and travel up through a preheater system which consists of a combination of a calciner and a set of cyclones. The raw materials consisting of a finely pulverized mix of Limestone and additive materials consisting of Iron, Silica, and Aluminium are sent to the top of the pre-heater tower from where they travel down in the kiln due to gravity. The hot gases traveling up through the pre-heater system heat up the raw materials where they are calcined first and then dropped in the kiln and then converted into a liquid phase in which the reaction takes place to produce the desired configuration of clinker

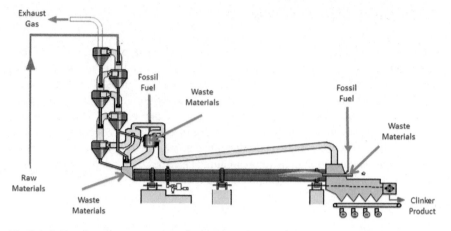

Fig. 9.1 Salient features of cement Kiln operation

chemistry. The liquid phase while travelling out of the kiln from the other end of the kiln gets cooled into solid clinker in a cooler and then sent out.

The objective of the co-processing operation is to replace the fossil fuels and raw materials with waste materials. The waste materials are therefore called Alternative Fuels and Raw materials (AFRs). These AFRs are fed in the cement kiln through different entry points as shown in Fig. 9.1. These entry points are termed as calciner, kiln burner, and kiln end. The AFR feeding can be done through any or all of these three entry points.

When the AFRs enter the kiln, its combustible fraction gets fully combusted as a fuel and the ash fraction gets mixed with the fossil material and gets utilized as raw material. The exhaust gases get released from the pre-heater top and are let out through the stack after sending it through the dedusting system such as ESP or Fabric filter.

9.4 Impact of Co-processing on the Emissions from the Cement Kiln

In the cement manufacturing process, the materials introduced in the kiln get subjected to very high temperature and long residence time. Under these conditions, the efficiency of combustion of the combustible matter is very high, which is much higher than that of the incinerator. This high level of combustion efficiency ensures a reduction in the VOC escaping out of the kiln system to the desired levels. This combustion efficiency is measured in terms of Destruction and Removal Efficiency (DRE).

The acid gases are generated due to the oxidation of elements such as Chlorine, Phosphorous, Florine, Sulphur, etc. They will, however, react with the hot calcined lime present in the preheater section and get neutralized fully, and hence they are not found in the exhaust gases at higher concentrations. Heavy metals present in the input materials entering the kiln system get embedded in the clinker matrix very strongly and remain in it as a non-leachable component forever. Only Nitrogen Oxides that get generated in the kiln system tend to get exhausted because there is no provision available in the kiln system to treat the same.

The constituents, that are responsible for the emissions in the kiln stack, are heavy metals, Chlorine, Phosphorous, Florine, Sulphur, difficult to burn organics, etc. The concentration of particulate emissions occurring in the kiln stack would depend upon the efficiency of the dust collection system. Hence, for lower emissions, high-efficiency systems are desired. The acidic gases are not a cause of concern as they get scrubbed with the lime in the preheater section and get neutralized.

One of the major hazards in the emissions in the flue gas from cement kilns utilizing AFRs are the heavy metal (HM) content. Not all heavy metals are toxic and not all toxic heavy metals have the same toxicity. Therefore, many countries differentiate between different toxicity classes:

- Class I: Cd, Hg, Tl;
- Class II: As, Co, Ni, Se, Te;
- Class III: Pb, Cr, Cu, Pt, V, Sn, Pd, Sb, Mn, Rh.

The heavy metal in class I are the most toxic and harmful, the heavy metal in class III the least ones. The main sources of heavy metal emissions from cement kiln stacks are either raw materials or fuels containing heavy metals (Jovovic et al., 2010). The quantum of heavy metals present in the flue gases would depend upon the volatility of the heavy metal or its compounds present in the kiln system. If the heavy metal content in the kiln system is more than the embedding capability of the clinker matrix, then the heavy metal would tend to get leached out from the clinker or the cement manufactured from this clinker.

Nitrogen Oxides are other emissions that are a cause of concern because they do not get treated in the kiln system unless a specific provision is made for the same. Usually, the control of Nitrogen Oxide is achieved using DeNOx systems. The most popular one in the cement industry is the selective Non-Catalytic Reduction (SNCR) System.

The VOC quantum present in the kiln flue gases would depend upon the efficiency of combustion. The lower the efficiency, the higher is the quantum. It is desired, therefore, that the combustion efficiency is improved using various means such as temperature, burner momentum, oxygen level, flame intensity, etc.

9.5 Monitoring of the Emissions from the Cement Kiln

Monitoring of emissions can be done offline or online. Offline monitoring involves the collection of the sample from the stack gas as per the defined methodology and analysis of the components present in the same. Online monitoring consists of measurements made by installing the instruments in the flue gas stream.

Different technologies are utilized for the measurement of the different emission constituents and many of the constituents can be measured these days online. The standard measurement principles associated with online (continuous) monitoring of different parameters are depicted in Table 9.1.

The major requirements of an efficient Continuous Emission Monitoring System (CEMS) are defined by the Central Pollution Control Board (CPCB) in India are mentioned below:

a. It should be capable of operating unattended over a prolonged period.
b. It should produce analytically valid results with precision/ repeatability.
c. The analyzer should be robust and rugged, for optimal operation under extreme environmental conditions, while maintaining its calibrated status. 1st Revised Guidelines for Continuous Emission Monitoring Systems August 2018 6.
d. The analyzer should have inbuilt zero check capability or external capability with a condition that no human intervention should be required to carry out a daily check at a defined time.
e. It should have a data validation facility with features to transmit raw and validated data to the central server at SPCB/CPCB. The data validation will be done after approval of SPCB/PCC or after 07 days of submission of a request for validation to SPCB/PCC wherever is earlier.
f. It should have Remote system access from the central server for provisional log file access. The facility shall be incorporated in the system within 06 months of the issue of these 1st Revised Guidelines.
g. It should have provision for simultaneous multi-server data transmission from each station without an intermediate PC or plant server.
h. It should have a provision to send system alarm to the central server in case any changes are made in configuration or calibration. The facility shall be incorporated in the system within 06 months of the issue of these 1st Revised Guidelines.
i. It should have a provision to record all operational information in a log file.
j. There should be provision for independent analysis, validation, calibration, and data transmission for each parameter.

Table 9.1 Principle of operation of emission measuring devices

Parameter	Principle
Particulate Matter	Opacity
Oxygen	ZrO_2
H_2O, CO_2, CO, N_2O, NO, NO_2, SO_2, HCl, HF, NH_3, CH_4, C_2H_6, C_3H_8, C_2H_4, C_6H_{14}, and CH_2O. Total Organic Carbon (TOC), SO_2, NOx, HCl, HF	Fourier Transform InfraRed (FTIR)
Hg	Cold Vapor Atomic Fluorescence (CVAF)

k. The instrument must have a provision of system memory (non-volatile) to record data for at least one year of continuous operation. Existing instruments not having adequate system memory shall be backed up with external devices within six months. All new instruments installed shall have an inbuilt provision of system memory.

l. It should have the provision of Plant level data viewing and retrieval with the selection of ethernet, Modbus, and USB.

m. Record of calibration and validation should be available on real-time basis at the central server from each location/parameter.

n. Record of online diagnostic features including analyzer status should be available in the database for user-friendly maintenance.

o. It must have low operation and maintenance requirements.

The CEMS system is also proposed to include the following features as per CPCB:

a. Continuous measurements on a 24 × 7 basis.

b. Direct Measurement of pollutant concentration.

c. Expression and display of measurements in ppm, mg/m^3 or volume % as specified in standards.

d. Display the measurement values as well as all the information required for checking/maintenance of the analyzer.

e. Display of functional parameters.

f. Response time <200 s.

g. Power supply compatible with utilities available on Indian industrial sites.

h. Digital communication with the distant computer for data acquisition/recording/reporting.

i. RS232/RS485/Ethernet/USB communication ports.

j. Analog Outputs for transmission to Plant's supervision centre.

k. The maximum lifetime of analyzers should be restricted to the expected life period specified by the Vendors or upon the perusal of deterioration in the performance of the analyzers i.e., frequent breakdowns and requirement of minimum data capture are not on it.

l. Type approved according to Indian Certification Scheme (or by foreign accredited institutes such as TÜV, MCERTS or USEPA).

9.6 Notified Emission Standards for the Cement Kilns in Different Countries

Ministry of Environment, Forest, and Climate Change (MoEFCC) of the Government of India has notified the following emission standards in 2016 for the cement kilns (MoEFCC, 2016). These standards are notified for both types of kilns—those which are not undertaking co-processing of AFRs and those which are undertaking co-processing of AFRs. MoEFCC has also mandated that cement plants monitor Particulate matter, SO2, and NOx online and transmit the values directly from the plant to the server of CPCB and relevant SPCB so that the same are available online assessment by the authorities. This mandate is for both kinds of plants—those that are co-processing and those that are not co-processing.

The co-processing plants must undertake the monitoring of all other mandated parameters such as VOC, Heavy metals, Dioxin and Furans, etc., through accredited analytical laboratories once a year and submit the results to CPCB and relevant SPCB. There are different accredited laboratories that are approved by the authorities to carry out the stack analysis to measure and report the emissions from the same to ascertain that the cement kilns are complying with the prescribed emission standards. These emissions could be both from co-processing and non-co-processing kilns. One such laboratory from India is VIMTA Laboratories Ltd. Its detailed features are illustrated in Fig. 9.2.

The prescribed emission limits in different countries (Edvards, 2014, Da Hai Yan et al., 2014, European union, 2020) are demonstrated in Table 9.2. The values in Table 9.2 have been grossly simplified to provide a broad overview of the prescribed emission standards in different countries. For a detailed understanding of the same, it would be desirable to review the actual notifications made by the respective countries. The main understandings from the table are that the emission standards in developed countries are more stringent than in developing countries.

VIMTA LABS LIMITED

(www.vimtalabs.com)

Vimta is the first environmental laboratory in India to be recognized under the Environment Protection Act 1986 by the Ministry of Environment and Forests, Government of India. Its experienced team carries out analysis of water, wastewater, soil, solid waste, hazardous waste as per Indian and international protocols for various parameters. It is also the first laboratory in India to establish an ultra-trace laboratory to test dioxins and furans in air emission, liquid waste, solid waste and food stuff, as per EPA standards with HR-GC/MS autospec, waters (the referral equipment for confirmatory testing of dioxins).

It has all the required accreditation and approvals & state of the art laboratories. Vimta's experience covers almost all the operating industrial sectors. It has carried out more than 1000 projects related to Characterization of Waste materials and impact assessment study on flue gas emission during AFR Co-processing consisting of Particulate Matter, Velocity, Moisture, Carbon Dioxide, Oxygen, Carbon Monoxide, Sulfur Dioxide, Nitrogen Oxides, HCl, HF, HBr, Volatile Organic Compounds, Benzene, Ammonia, Heavy Metals, Mercury, PAH, Dioxin & Furan etc.

Vimta's experience covers a wide range of sectors including power, chemical, cement, mining, steel & alloys, metallurgical, dye & intermediates, bulk drugs, pesticides, agrochemicals, petro chemicals, refineries, pulp & paper, oil & gas exploration & production, oil & gas pipelines, foundries, airports, sea ports, jetties, industrial parks/ SEZs, building/ infrastructure etc.

Fig. 9.2   Features of an emission testing laboratory—VIMTA Labs Ltd. India

Table 9.2 Prescribed emission limits in different countries (*Source* Edvards, 2014)

Parameter	PM		SOx		NOx		Hg
Country	mg / Nm^3		mg / Nm^3		mg / Nm^3		mg / Nm^3
	Case1	Case2	Case1	Case2	Case 1	Case2	
Germany	20		50		200		0.03
Austria	20		350		500		0.05
EU	30		50		500	800	0.05
China	20	30	100	200	400	320	0.05
India	30		100		600	800	0.05
UK	30		200		500		0.05
Australia (New South Wales)	95		50		800		0.1
Egypt	100	50	400		600		0.05
Nigeria	100		2000		1200		0.05
Turkey	120	50	300		400		None
Columbia	150	250	500		550		None
Bolivia	300		600		1800		None

Table 9.3 Co-processing specific emission standards

Country	TOC	HCl	HF	Cd + Tl	Heavy Metals	Dioxin & Furan
Units	Mg / Nm^3	Mg / Nm^3	Mg / Nm^3	Mg / Nm^3	Mg / Nm^3	ng TEQ/Nm^3
EU	10	10	1	0.05	0.5	0.1
China	10	10	1	1.0*	0.5	0.1
India	10	10	5	0.05	0.5	0.1

*Along with Pb and As

In some countries there are graded standards to accommodate certain situations/conditions such as quality of raw materials & fuel, age of the plant, adaptability of the available technology, etc. in addition to the above standards, many countries have also co-processing specific standards. These are provided in Table 9.3.

9.7 Conclusion

Different kinds of emissions are caused when a material is thermally treated. In cement kilns also different kinds of emissions take place. Some of them get trapped due to the alkaline environment and some tend to exit along with the exhaust gases. These emissions have different impacts on the environment and also on the living beings and hence need to be properly monitored and controlled. Different countries have notified limiting values for these emissions for the cement kilns in the country to comply with. In this chapter, different emissions caused in the cement kiln, their impacts, monitoring, and control mechanisms, prescribed standards by different countries to comply while undertaking co-processing, and otherwise have been discussed.

References

Adeyanju, E., & Okeke, C. A. (2019). Exposure effect to cement dust pollution: A mini review. *SN Applied Science, 1*, 1572. https://doi.org/10.1007/s42452-019-1583-0.

CPCB. (2010). Guidelines on co-processing in Cement/Power/Steel Industry, published by CPCB in Feb 2010.

Edvards, P. (2014). Global cement emissions standards. *Global Cement Magazine,* 24–33

European Union Directive 2000/76/EC, IPCC, 2013; Council Directive, 2000

GSR 497 E (2016). Moefcc, Government of India, notification dated 10th May 2016

Gupta, R. K et al. (2012). Particulate matter and elemental emissions from a cement kiln. *Fuel Processing Technology, 104*, 343–351. https://www.researchgate.net/publication/257210587_Particulate_matter_and_elemental_emissions_from_a_cement_kiln

Jovovic, A. et al. (2010). The emission of particulate matters and heavy metals from cement kilns-case study: Co-incineration of tires in Serbia. *Chemical Industry and Chemical Engineering Quarterly, 16*(3). https://doi.org/10.2298/CICEQ090902010J

Karstensen, K. H. (2008). Formation, release and control of dioxins in cement kilns. *Chemosphere, 70*(4), 543–60. https://doi.org/10.1016/j.chemosphere.2007.06.081. Epub 2007 Aug 14. PMID: 17698165 Review.

Khattak, Z., et al. (2013). Contemporary dust control techniques in cement industry, electrostatic precipitator - a case study. *World Applied Sciences Journal, 22*(2), 202–209.

Lanids Wayne, G. et al. (2000). *Introduction to Environmental Toxicology: Molecular Substructures to Ecological Landscapes*, Fifth Edition. CRC Press

Liu, G., Zhan, J., Zheng, M., Li, L., Li, C., Jiang, X., Wang, M., Zhao, Y., Jin, R., & Hazar, J. (2015). Field pilot study on emissions, formations and distributions of PCDD/Fs from cement kiln co-processing fly ash from municipal solid waste incinerations. *Mater, 15*(299), 471–478. https://doi.org/10.1016/j.jhazmat.2015.07.052 Epub 2015 Jul 26PMID: 26241773.

Liu, G., Zhan, J., Zhao, Y., Li, L., Jiang, X., Fu, J., Li, C., & Zheng, M. (2016). Distributions, profiles and formation mechanisms of polychlorinated naphthalenes in cement kilns co-processing municipal waste incinerator fly ash. *Chemosphere, 155*, 348–357. https://doi.org/10.1016/j.chemosphere.2016.04.069 Epub 2016 Apr 29PMID: 27135696.

Manisalidis, I. et al. (2020). Environmental and health impacts of air pollution: A review. *Front Public Health, 8*, 14. Published online 2020 Feb 20. https://doi.org/10.3389/fpubh.2020.00014

Mukherjee, A., Debnath, B., Ghosh, S. K. (2016). A review on technologies of removal of dioxins and furans from incinerator flue gas. *Procedia Environmental Sciences, 35*, 528–540, ISSN 1878-0296, https://doi.org/10.1016/j.proenv.2016.07.037. (http://www.sciencedirect.com/science/article/pii/S1878029616301268)

Shah, K., An, N., & Ma, W. et al. (2020). Chronic cement dust load induce novel damages in foliage and buds of Malus domestica. *Science Reports, 10*, 12186. https://doi.org/10.1038/s41598-020-68902-6

Shibamoto, T., Yasuhara, A., & Katami, T. (2007). Dioxin formation from waste incineration. *Reviews of Environmental Contamination and Toxicology, 190*, 1–41. https://doi.org/10.1007/978-0-387-36903-7_1.PMID:17432330Review

Stephania, A. et al. (2006). Origin and health impacts of emissions of toxic by-products and fine particles from combustion and thermal treatment of hazardous wastes and materials. *Environmental Health Perspectives, 114*(6), 810–817. Published online 2006 Jan 26. https://doi.org/10.1289/ehp.8629

Tuomisto, J. (2019). Dioxins and dioxin-like compounds: Toxicity in humans and animals, sources, and behaviour in the environment. *WikiJournal of Medicine, 6*(1), 8. https://doi.org/10.15347/WJM/2019.008

Yan, D., et al. (2014). China's new emission limits. *International Cement Review.*

Yang, L., Zheng, M., Zhao, Y., Yang, Y., Li, C., & Liu, G. (2019). Unintentional persistent organic pollutants in cement kilns co-processing solid wastes. *Ecotoxicology and Environmental Safety, 182*, 109373. https://doi.org/10.1016/j.ecoenv.2019.109373. Epub 2019 Jun 28. PMID: 31255869.

Zou, L., Ni, Y., Gao, Y., Tang, F., Jin, J., & Chen, J. (2018). Spatial variation of PCDD/F and PCB emissions and their composition profiles in stack flue gas from the typical cement plants in China. *Chemosphere, 195*, 491–497. https://doi.org/10.1016/j.chemosphere.2017.12.114 Epub 2017 Dec 19 PMID: 29274995.

Chapter 10
Co-processing of Wastes as AFRs in Cement Kilns

10.1 Introduction

Co-processing is a thermal treatment process to utilize wastes as resources in the Resource Intensive Industries (RIIs) such as Cement kilns, Thermal Power Plants, Steel Plants, Glass Manufacturing plants, Refractory Manufacture plants, and Lime plants. Wastes get utilized as Alternative Fuels and Raw materials (AFRs) in the RIIs and replace the fossil fuels and raw materials used traditionally in the RIIs. Among the different RIIs, the cement manufacturing process has special features that allow large quantum usage of wastes as AFRs in the cement kiln. When wastes get utilized as AFRs, they replace the use of fossil fuels and fossil raw materials thereby conserving them for future use. These special features of the kiln also provide opportunities for wastes to get sustainably managed in the cement manufacturing process without impacting the quality of the clinker. Clinker is an intermediate product in the manufacture of cement.

10.2 Co-processing of Wastes as AFRs in the Cement Kilns

Cement kiln co-processing technology provides opportunities for the sustainable management of wastes from Municipal, Industrial, and Agricultural sectors. These wastes have organic content and inorganic content. Organic content is a combustible material, and it gets combusted during co-processing releasing the energy present in it. The release of this energy in the kiln causes reduction in the fossil fuel consumption of the kiln. The inorganic content present in the waste stream is generally a mixture of oxides of Calcium, Iron, Silicon, Aluminium, and other metals. The raw materials used in cement manufacture are also a mix of similar types of metal oxides. Hence, this inorganic content gets utilized as raw material in cement manufacture. Therefore, the waste material gets utilized as alternative fuels and raw materials (AFRs) in the cement kiln co-processing.

© The Author(s), under exclusive license to Springer Nature Singapore Pte Ltd. 2022   197
S. K. Ghosh et al., *Sustainable Management of Wastes Through Co-processing*,
https://doi.org/10.1007/978-981-16-6073-3_10

10.3 Cement Kiln Operation

Cement manufacture consists of reacting oxides of Calcium, Iron, Silica, and Aluminium in certain proportions to produce an intermediate product called clinker. This reaction takes place at a high temperature of > 1400 °C in the cement kiln. The desired temperature in the kiln is achieved by firing appropriate fossil fuel in the kiln. Different kinds of fossil fuels used by the cement industry include coal, oil, or gas. A large amount of petcoke is also utilized by the cement industry as a fuel source. The raw materials fed in the kiln get heated up to calcination temperature (850–920 °C) in which the limestone gets converted into lime and then the raw material gets heated to temperatures up to 1450 °C when the same react with one another to form clinker having the desired characteristics. Clinker is an intermediate product in cement manufacture. This intermediate product, manufactured in the kiln, is then ground with gypsum to produce ordinary portland cement (OPC), with gypsum and fly ash to produce Portland pozzolana cement (PPC), with granulated blast furnace slag (GBFS) and gypsum to produce Portland slag cement (PSC) and with fly ash, GBFS, limestone, and gypsum to produce Portland composite cement (PCC).

Clinker is manufactured in two kinds of kilns—rotary kiln and vertical shaft kiln (VSK). The majority of the clinker gets manufactured in India and worldwide using the Rotary kiln process. In the earlier days, cement manufacture used to be carried out using wet process technology in which raw materials used to be converted in the form of slurry in water, and the slurry was fired in the kiln. This operation was being carried out in long wet kilns. Subsequently, the process changed to dry process technology in which the raw material used to get processed in the dry form to produce desired quality clinker. Later the long kiln technology was transformed into preheater-based kilns. In the long wet kiln, long dry kiln, and the preheater kilns, the fuel is fired in the main burner installed in the kiln. Now the latest technology is to utilize pre-calciner kilns. In these kilns, fuel is fired in the pre-calciner to calcine the raw material and then the raw material enters the kiln where it reacts to form clinker using the heat provided by the fuel that is fired in the kiln burner. There are two kinds of pre-calciner designs. The first one is in line calciner (ILC) and the other is separate line calciner (SLC). Currently, the majority of the newly installed kilns are designed with ILC.

10.4 Important Aspects in Respect of Co-processing

There are a few important aspects in respect of co-processing, and these are defined below:

10.4.1 TSR

TSR stands for Thermal Substitution Rate. It signifies the percentage of energy utilized from waste and AFR in the clinker in the manufacturing process. It is calculated using the following equation:

$$\textbf{TSR}\% = \left\{ \frac{\{\text{Thermal energy utilized from waste or AFR}\}}{\{\text{Total Energy utilized in the kiln process}\}} \right\} \times 100$$

10.4.2 AFR/Waste Profile

AFR/Waste materials do not have prepared material safety data sheets (MSDS). Therefore, it is desired that a similar sheet is generated for each AFR/waste stream that will provide a reasonably good understanding of its safe handling and processing. This is called AFR/Waste Profile. This needs to be generated by compiling the relevant information through interactions with the relevant stakeholders and doing the appropriate research (GTZ, 2006).

10.5 Salient Features of Cement Kiln Co-processing Technology

The diagram of the present-day pre-calciner kiln with relevant infrastructure is depicted in Fig. 10.1 below.

In Fig. 10.1, the brown cylinder with red flame is the cement kiln, the cyclone string is the preheater column, brown equipment with the red flame is the pre-calciner, blue equipment is the cooler, and the green equipment is the feed bin.

The raw material is fed from the feed bin into the kiln system. It travels down from the top cyclone to the bottom cyclone of the preheater tower and then travels inside the kiln. It gets heated in a counter-current manner by the hot gases that are travelling up from the kiln burner end and calciner to the top of the preheater tower. After exchanging heat with incoming material, the cooled gas gets exhausted into the atmosphere after dedusting it in an ESP or bag filter. The clinker formed in the kiln travels into the cooler and gets discharged after getting cooled.

The temperature of the flame in the kiln is > 1800 °C, and the temperature in the kiln at the other end is about 1100 °C. The temperature in the calciner is about 900 °C. The clinker production capacity of the kilns implemented at present is about 8000 TPD–12,000 TPD. This clinker production requires about 12,000 TPD–18,000 TPD of raw material and about 1000 TPD–2400 TPD of fuel depending upon its calorific value. The diameter of the kiln varies from about 4.5 to 6 M and the length of the kiln varies from 60 to 100 M. The height of the preheater tower ranges up to

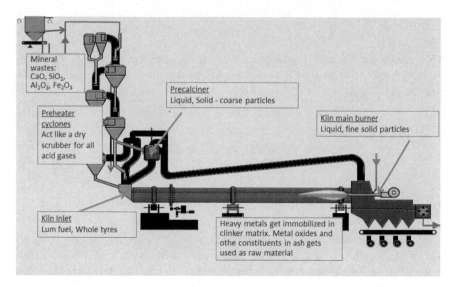

Fig. 10.1 Cement kiln features and material feed points for co-processing

about 180 M. Both raw material and fuel are ground to a very fine size that is less than 100 microns.

For co-processing, the feed point for introducing waste into the kiln system depends upon the characteristics of waste material and are depicted in Fig. 10.1. The different characteristics of the waste are the following:

- Lumpy waste: Tyres, waste-filled drums, and waste-filled bags. These are fed at the kiln inlet.
- Coarse solid waste: coarsely shredded plastic and RDF material, and solid chunks. These are fed in the calciner.
- Fine solid waste: Finely shredded plastics, and RDF and powdery material are co-processed through the main burner.
- Liquid wastes: Aqueous and organic liquids. These are fed through the pre-calciner and main burner [atomization required].

The salient features of the kiln that facilitate sustainable management of wastes are the following.

10.5.1   Zero Waste Technology

The organic content in the waste gets completely combusted and inorganic content becomes part of the raw material and reacts to become product clinker. Hence, co-processing is a zero waste technology.

10.5.2 Kiln Emissions are Not Influenced

Very high temperature (900 to > 1800 °C) and long residence time (6–10 s) available in the kiln ensures the organic content in the waste is fully combusted. The Destruction and Removal Efficiency of the kiln system is more than 100 times greater than that of an incinerator. Hence, there is no cause of concern as far as volatile organic compounds (VOC) are concerned.

10.5.3 Acidic Gases Get Absorbed in the Calcined Lime

Chlorine, fluorine, and sulphur, when subjected to thermal treatment, get converted into acid gases, namely HCl, HF, SO_2, and SO_3. These acid gases, while travelling up through the kiln systems, react fully with lime and get converted into Calcium Chloride, Calcium Fluoride, and Calcium Sulphate. Hence, these acid gases are not observed in the kiln emissions.

10.5.4 Dioxins and Furans

International studies generally indicate that most modern cement kilns today can meet an emission level of 0.1 ng I-TEQ/m3 and that proper and responsible use of organic hazardous and other wastes to replace parts of the fossil fuel is not an important factor influencing the formation of PCDD/PCDFs. Modern preheater/precalciner kilns generally seem to have lower emissions than older wet process or long dry cement kilns without a preheater, but the main influencing parameter stimulating the formation of PCDD/PCDFs seems to be the availability of organics/precursors in the raw material and the temperature of the air pollution control device. Feeding of materials containing elevated concentrations of organics as part of raw material mix should therefore be avoided and the exhaust gases should be cooled quickly in wet or long cement kilns.

10.5.5 Heavy Metals Get Fixed in the Clinker

Heavy metals present in waste get immobilized in the clinker matrix in such a manner that they do not get leached in water. However, there is a limit to the capacity the clinker has to embed the heavy metals in its matrix.

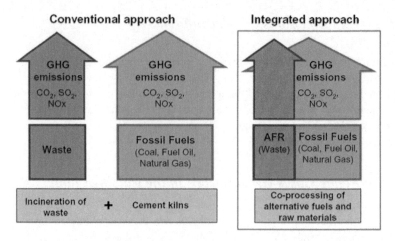

Fig. 10.2   Co-processing reduces GHG emissions

10.5.6   High Efficiency in Material and Energy Recycling

In co-processing, 100% of the material and 100% of energy present in waste gets recycled in the manufacture of clinker. One Kg of raw material present in waste [oxide content] replaces 1 kg of natural raw material and I KJ of energy coming from waste replaces 1 kJ of energy coming from the fossil materials.

10.5.7   Reduction in Global Emissions

The quantum of emissions from cement kiln does not change due to the co-processing of waste materials in place of fossil materials. Since co-processing avoids disposal or degradation of the waste materials, the corresponding emissions also get avoided. Hence at a global scale, the emissions get reduced. This is depicted in Fig. 10.2.

10.5.8   Local and Cheaper Solution for the Local Problem

Cement industries that are located close to waste generation locations can provide the sustainable solution of co-processing locally. Further, the cement plants are already present and hence the investment required to set up waste management infrastructure is substantially less compared to a new solution that needs to be implemented. Further, since the waste is replacing the natural resources, the economic proposition is more favourable compared to options such as Waste to Energy and Incineration.

10.6 Different Kinds of Wastes Suitable for Co-processing

All wastes derived from Municipal, Industrial, and Agricultural sectors are suitable for co-processing except the banned items. The following is the list of different wastes that can be co-processed in cement kilns.

10.6.1 Municipal Sector

There are three kinds of waste that are derived out of MSW which can be utilized as AFRs in the cement kilns.

10.6.1.1 Non-Recyclable Plastic Waste

Non-recyclable plastic waste includes single-use plastic, multi-layer packaging, contaminated plastic scrap, thermo-set plastic waste, etc. Generally, this kind of plastic waste is available from (a) agencies who are recycling plastic waste, (b) brands who have an obligation under extended producer responsibility (EPR), (c) material recovery facilities of municipalities, etc. Non-recyclable Plastic Waste in bailed form is shown in Fig. 10.3.

Fig. 10.3 Non-recyclable plastic waste

Fig. 10.4 SCF from MSW

10.6.1.2   Segregated Combustible Fraction (SCF)

SCF from MSW is a mix of waste plastics, paper, rexine pieces, rubber, shoes, chappals, tyre pieces, tubes pieces, old and torn clothes, contaminated plastics, paper and clothes, wood pieces, coconut shells, other biomass, etc. SCF is generally available from agencies who are processing mic garbage in an integrated MSW treatment facility, agencies who are processing a dry fraction of MSW, and agencies who are implementing dump yard remedying. SCF from MSW is shown in Fig. 10.4.

10.6.1.3   Dried Sewage Sludge (DSS)

DSS is generated during the treatment of the sewage sludge. The desired moisture content in DSS for use as AFR in cement kiln is less than 10%. Dried Sewage Sludge is shown in Fig. 10.5.

Fig. 10.5 Dried sewage sludge

10.6.2 Industrial Sector

Hazardous as well as non-hazardous wastes are generated during industrial manufacturing processes. These can be in the form of solid, liquid, and sludges. Hazardous wastes consist of date-expired medicines, ETP sludges, distillation residues, process residues, paint sludges, chemical sludges, obsolete pesticides, failed batches containing hazardous substances, etc. Non-hazardous wastes consist of packaging items, date-expired products from FMCG and food sector, etc.

These are demonstrated in Figs. 10.6 and 10.7 respectively.

10.6.3 Agricultural Sector

This includes non-cattle feed bio-waste such as agricultural waste, biomass, tree/leaves felling, pulse stems, crop residue, residue from food processing industries, etc.

Figure 10.8 depicts a picture of agro-wastes.

Fig. 10.6 Hazardous waste

Fig. 10.7 Non-hazardous waste

Fig. 10.8 Agro-waste

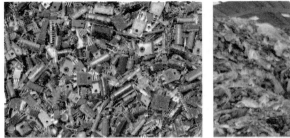

Electronic Waste

Unsorted municipal
solid waste

Entire Batteries

Fig. 10.9 Items that can be co-processed only after pre-processing

10.7 Items Which Cannot/Should Not Be Co-processed

The following items are banned for pre-processing and/or co-processing from various
safety considerations.

10.7.1 Items that Can Be Co-Processed Only After Pre-processing

Items that can be co-processed only after pre-processing are shown in Fig. 10.9.

10.7.2 Items that Cannot Be Pre-Processed or Co-processed

Items that cannot be pre-processed or co-processed are shown in Fig. 10.10.

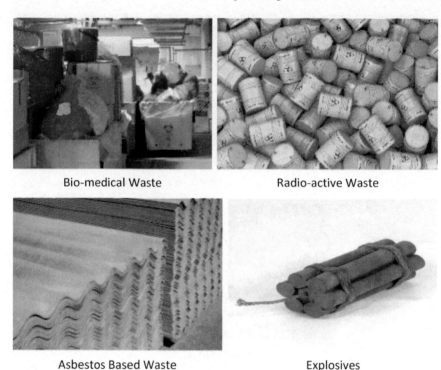

Bio-medical Waste   Radio-active Waste

Asbestos Based Waste   Explosives

Fig. 10.10   Items that cannot be pre-processed or co-processed

10.8   Typical Examples of AFRs and Their Sources

As per the Basal Convention, co-processing is recommended for the safe management of hazardous wastes and POPs (Basel Convention, 2011). As per the guidelines published by CPCB (Guidelines, 2017)of India, all kinds of hazardous wastes, SCF, RDF, plastic and other packaging wastes, tyre chips, non-hazardous industrial wastes, biomasses, agro-wastes, date-expired or off-specification FMCG, food and kindred and other products, etc., are feasible to be co-processed in the cement kilns. As per CPCB, co-processing of wastes can be implemented while complying with the prescribed emissions standards.

A list of typical wastes/AFRs which can be co-processed in cement kilns is provided in Table 10.1 along with the feasible agencies from whom they can be sourced.

Table 10.1 Wastes that can be co-processed and their sources

S. No.	Waste	Source
1	Biomass/Agro-waste	Local farming community, forest department, grain processing industries, food processing industries, wood processing industries, municipalities, traders dealing with biomass supply, etc.
2	SCF	Material Recovery Facilities created by municipalities, Integrated Waste Management Facilities, dump yard remedying project sites, NGOs, EPR projects implemented by PIBOs
3	RDF	SCF to RDF processing plants, plastic pyrolysis projects
4	Plastic waste	Plastic recyclers, municipalities, PIBOs implementing EPR projects
5	Tyre Chips	Tyre manufacturers, tyre processing agencies, rubber crumb manufacturing industries, pyrolysis plants, imports
6	Rubber waste	Rubber products manufacturing industries, shoe sole manufacturing companies, shoe manufacturing companies
7	Hazardous wastes	TSDF operators, industries generating hazardous wastes, hazardous waste processing industries, FRP and SMC products manufacturing companies, pharmaceutical companies for date-expired medicines, etc.
7	Non-hazardous Wastes	Non-hazardous waste generating industries, tyre pyrolysis projects, sponge iron industries, FMCG, food and kindred companies, water treatment projects

10.9 Important Considerations for Smooth and Successful Co-processing

10.9.1 Pre-Processing of Wastes into AFRs

Generally, wastes do not have quality specifications. They vary considerably from lot to lot and source to source. In cement kiln, uniformity in the quality of all input materials is very important from the point of view of smooth kiln operation and clinker quality. Hence, for achieving the desired quality of AFR from waste streams, their pre-processing is required. For co-processing higher levels of AFRs, the standard deviation in the quality of pre-processed AFR needs to be lowered.

Pre-processing can be carried out either by a cement plant or by a third-party agency depending upon the convenience of the relevant business model.

10.9.2 Monitoring and Control of AFRs While Co-processing

AFRs are derived out of waste materials and may therefore contain constituents that can influence emissions, impact product quality and disturb the clinkerization process. Hence, it is important to monitor the AFR quality in required detail to ensure smooth operation of the cement kiln. The following are the important quality parameters that need to be evaluated for smooth and trouble-free cement kiln co-processing operation:

- Net calorific value;
- Ash content;
- Moisture content;
- Chlorine content;
- Sulphur content;
- Ash analysis in terms of Cao, SiO_2, Al_2O_3, FeO, and other oxides;
- Heavy metal content;
- Viscosity of liquid waste stream;
- Unacceptable constituent present in liquid waste streams;
- Flash point.

Therefore, an important aspect is to also set up a laboratory with suitable instruments at the co-processing site.

10.9.3 Emission Monitoring During Co-processing

As AFRs are derived out of waste materials and everything present in waste may not have been analysed (for example, heavy metals and complete Ash analysis), it is important to monitor the emissions from the kiln on real-time basis using Continuous Emission Monitoring System (CEMS).

Currently, CEMS with different design considerations are available for installation on the cement kiln stack. Some of these provide an online assessment of individual parameters such as dust, SOx, NOx, and VOC. There are others which are more versatile and in a single probe can provide online monitoring of various parameters simultaneously such as SOx, NOx, H_2O, HCl, HF, CO_2, CO, VOC, and NH_3.

The statutory provision in India demands that the output from the online CEMS installed on the kiln stack measuring dust, Sox, and NOx is connected to the data servers of the pollution control boards of state and the centre.

10.9.4 Permissions for Co-processing of Wastes

Co-processing operation involves utilizing wastes as resources. Wastes have liabilities and are uncertain in their behaviours. Therefore, as per regulatory frameworks of the respective countries, the pre-processing and co-processing of wastes get monitored and controlled by authorities who have been assigned the relevant responsibilities of granting permissions. These permissions are of two kinds: (1) Setting up facilities for pre-processing and co-processing of wastes and AFRs and (2) for handling, storage, pre-processing, and co-processing of the waste materials in these facilities.

In India, appropriate regulation related to wastes is defined by Ministry of Environment, Forest and Climate Change and monitored at the centre by Central Pollution Control Board (CPCB) and at the state level by the respective State Pollution Control Boards (SPCBs). For the agency that intends to generate, handle, transport, process, manage, and dispose of any kind of waste, specific permissions are required to be obtained by this agency from appropriate authority through the defined process.

Co-processing of wastes as AFRs in cement kilns also needs to be performed after obtaining necessary permissions which will be granted by the CPCB and/or SPCBs as per the provisions made in the respective rules. In India, the governing rules are Solid Waste Management Rules 2016 (SWM Rules 2016), Hazardous and Other Waste Management Rules 2016 (HOWM Rules), Plastic Waste Management Rules 2016 (PWM Rules 2016), or the subsequent amendments/revisions made in these rules.

Co-processing operation also requires an infrastructural facility to pre-process and co-process the wastes. Permission is also required for the installation and operation of this infrastructure. This is also given by CPCB and respective SPCB as per the provisions of the rules.

Permissions for co-processing are also required when some specific materials are proposed to be co-processed such as persistent organic pollutants (POPs) and ozone depleting substances (ODS). These permissions are granted based on the results of the co-processing trial which needs to be carried out as per the defined procedure.

Cement kiln co-processing is a globally practiced technology for the management of different kinds of wastes in an environmentally sound and ecologically sustaining manner. Different types of wastes, e.g., agro-wastes, hazardous and non-hazardous industrial wastes, and segregated combustible fractions (SCF) from MSW are disposed of in large quantities in many cement kilns in different countries all over the world.

In India, cement plants have implemented about 90 co-processing trials of wastes to prove the acceptability of cement kilns for their environmentally sound disposal. Cement plants have implemented these demonstration trials on different kinds of waste streams as per the protocol prescribed by Central Pollution Control Board. Based on the critical evaluation of the results of these trials, Central Pollution Control Board have endorsed the acceptability for their co-processing in cement kilns.

A critical review of the results of such 22 co-processing trials—which were endorsed as successful by CPCB—were evaluated to understand the extent of variation present in the chemical constituents of these waste streams. It was concluded through this evaluation that different waste streams having large variations in chemical characteristics can be managed in an environmentally sound manner in the cement kilns. Variation in the quantum content of different constituents present in the waste streams utilized in these 22 trials is provided in Table 10.2 below (Ulhas Parlikar et al., 2016).

The important aspect of these trials was that the quantum of AFR utilized in these different trials was different in the respective trials.

Table 10.2 Composition of
various constituents in the
waste materials co-processed
in cement kilns

PARAMETER	MIN	MAX
Moisture (%)	0.60	67.4
Ash (%)	0.96	98.70
Volatile Matter (%)	0.3	94.9
Fixed Carbon (%)	0.1	45.7
Carbon	0.4	75.6
Hydrogen	0.2	9.1
Nitrogen	0	15.5
Sulphur	0.1	22
Oxygen	0	76.3
Gross Calorific Value (Kcal/Kg)	80	7960
Net Calorific Value (Kcal/Kg)	114.8	6042
Mineral Matter	3.5	34.5
Chloride as Cl (mg/kg)	0	14,200
Fluoride as F (mg/kg)	0	20.1
Moisture (%)	0.60	67.4
Ash (%)	0.96	98.70
VM (%)	0.3	94.9
FC (%)	0.1	45.7
Carbon	0.4	75.6
Hydrogen	0.2	9.1
Nitrogen	0	15.5
Sulphur	0.1	22
Oxygen	0	76.3
GCV (Kcal/Kg)	80	7960
NCV (Kcal/Kg)	114.8	6042
Mineral Matter	3.5	34.5
Chloride as Cl (mg/kg)	0	14,200
Fluoride as F (mg/kg)	0	20.1
VOC (mg/kg)	4.20	207.0
SVOC (mg/kg)	BDL	0.2
PCB (mg/kg)	0.00	0.5
PCP (mg/Kg)	BDL	1.4
TOC (%)	0.00	66.0

Based on the results of these 90 trials, co-processing technology has been recognized in the legal framework of India and also guidelines have been published which define the procedure for permitting co-processing of hazardous and non-hazardous wastes.

As per the rules notified in 2016, permission for co-processing can be granted for different waste streams—except POPs and ODS materials—without undertaking any co-processing trials.

10.10 Principles of Co-processing

The following are the principles that need to be followed for responsible co-processing so that it contributes to the sustainable management of wastes.

Principle I—Co-processing shall not be used in cement kilns if better ecological ways of recovery are available.

Principle II—During AFR co-processing, emissions from the cement kiln shall not be higher than those with traditional fuel.

Principle III—The clinker, cement, and concrete shall not be abused as a sink for heavy metals.

Principle IV—Companies and personnel engaged in co-processing should have good environmental and safety compliance track records.

Principle V—Implementation of co-processing has to take into consideration country-specific regulations and procedures.

Principal VI—Availability of proper infrastructure to control AFR feed rate.

There are various guidelines that are practiced internationally to ensure that wastes are managed in an environmentally sound and safe manner. These guidelines and practices are required to be practiced suitably while undertaking pre-processing and co-processing of wastes/AFRs. These guidelines and practices are elaborated in detail in Chap. 6 of this book.

10.11 Co-processing Technology

AFR Co-processing technology consists of different steps that are explained in detail below.

10.11.1   Receipt of Waste/AFR Material

At the cement plant undertaking co-processing, depending upon the co-processing rate, either waste or AFR will be received. The amount of waste that can be fed in the kiln without impacting the cement kiln process or quality of the clinker product would depend upon the nature and quality of the waste stream.

However, generally, if the utilization quantity is < 5% TSR, waste as such may be acceptable to be fed in the kiln. At this feed rate, co-processing will proceed without any major concerns. However, if the quantum utilization is > 5% TSR, pre-processing of waste to AFR, meeting required specifications, would need to be implemented. This pre-processing may be implemented by the cement plant or also by a third party. The received waste/AFR is then taken inside the store after weighment if the quality of the same is seen as acceptable/as per the agreement.

10.11.2   Quality Assessment of Incoming Waste/AFR Material

The laboratory is an important consideration in undertaking co-processing because quality assessment of the waste/AFR stream has a substantial influence on the clinker quality, emissions, and process of operation. The parameters to be evaluated depend upon the nature and physico-chemical characteristics of the waste/AFR stream. Some of the important considerations of the ingredients present in wastes/AFRs, which need to be understood properly while co-processing, are the following:

1.   **Ash content:** When the quantum of AFR usage starts increasing, the composition of ash becomes an important parameter because it may influence the raw mix design. Hence, ash analysis of the wastes/AFRs needs to be done and needs to be taken into consideration while designing the raw mix.
2.   **Chlorine content:** Chlorine is an important parameter in the cement process as there is a limit up to which it is permitted in the cement product and also it influences the process by causing a coating of the alkali chlorides on the surfaces reducing the flow path of gas and increasing pressure drop in the system.
3.   **Phosphorous content:** Phosphorous, when present in the raw mix beyond a certain proportion, has a tendency to reduce the setting time of the cement and needs to be critically evaluated.
4.   **Fluorine content:** Fluorine acts as a mineralizer and reduces the liquid formation temperature in the clinker manufacturing process.
5.   **Presence of heavy metals:** The presence of heavy metals in wastes/AFRs is a matter of concern in co-processing due to toxicity and other considerations. There are different guideline limits that have been specified in the waste/AFR streams (GTZ, 2006).

a. **Chromium content:** Chromium in cement in (+6) form tends to get leached out easily when cement comes in contact with water. Chromium also causes severe skin irritation. Its excessive content in wastes/AFRs is undesired.

b. **Mercury, Cadmium, and Thallium content:** The compounds of these metals are volatile and highly toxic. Their excessive presence in the wastes may cause the operating temperature in the clinkerization process and tend to get released as toxic emissions. Their excessive content in wastes/AFRs is undesired.

c. **Present of other heavy metals:** The crystalline phases in clinker have the ability to embed heavy metals into them and make them non-leachable. However, there is a limit for the same to happen in the clinker matric. Beyond this value, the heavy metals tend to get leached out causing environmental concern.

Depending upon the nature and source of the waste streams, the evaluations of the above parameters need to be taken up. For this, appropriate instrumentation systems need to be set up in the laboratory.

The design of the laboratory needs to be properly laid out so that safety prevails. Generally, the facilities like oven furnace, etc., are kept in a separate room so that the exhaust gases are properly vented out. Sophisticated instruments such as XRF, ICP-MS, and GC need to be installed in closed and air-conditioned rooms. Laboratory also needs to have a proper security system to preserve the safety of the facility, safety of the data, and its integrity.

The laboratory also requires a sample preparation facility and sample storage facility. The samples need to be preserved for a defined duration depending upon whether the sample is a pre-qualification sample or the standard delivery sample. This requirement of preserving of a sample is very important from the point of view of traceability and auditability in case it is demanded in the given case. Usually, the pre-qualification sample is preserved for 6 months to 1 year and the standard delivery sample, which is evaluated for fingerprint and detailed analysis, for a period of 1–3 months.

There are three kinds of sample testing involved in the case of wastes/AFRs.

a. **Pre-qualification analysis:**

This consists of evaluating the nature and physico-chemical characteristics of the waste/AFR stream for the first time to have it as a benchmark assessment for future reference and considerations. This analysis consists of ultimate and proximate analysis, halogen analysis, ash analysis, heavy metal content, and flash point. In case of liquid, the analysis is to be carried out by Gas Chromatograph.

b. **Fingerprint analysis**

Fingerprint analysis is carried out after the material is reached at the plant premises. This analysis is carried out to evaluate its results with the agreed quality considerations with the supplier and giving its acceptance for shifting it to plant stores. This analysis is carried out on a representative sample of the received material while the material is awaiting delivery at the gate.

This analysis is generally carried out to evaluate parameters that define the waste/AFR stream characteristics as agreed with the supplier. For Alternative Fuels, it would be GCV, moisture, chlorine, flash point, and ash content and for Alternative Raw materials, it will involve desired material content, moisture, chloride and sulphur content, etc. If the waste stream is liquid, then the GC analysis of the waste stream is carried out to compare the constituents with those in the reference sample.

c. **Detailed analysis**

A detailed analysis of the waste/AFR stream is carried out to plan out its pre-processing and co-processing actions. Usually, it involves preparing a representative sample from the total lot under consideration and analysing it in detail. This consists of ultimate analysis, proximate analysis, heavy metal content, ash content and the ash analysis, flash point, etc.

After the laboratory evaluation of the incoming material awaiting delivery at the gate, the same is cleared for delivery.

d. **Compatibility assessment of waste mixes:** Waste/AFR materials tend to contain materials having unknown characteristics. Sometimes, the mix of such waste/AFR streams of different origins may lead to heating, runaway reactions, polymerization, fires, and explosion. To ensure that these undesired reactions do not occur, their compatibility evaluation is desired to be carried out.

  a. **General evaluation:** The chemical compatibility chart depicted Fig. 6.4 in Chap. 6 needs to be referred to while storing the materials in the laboratory or in the waste stores. Incompatible materials need to be kept sufficiently away from each other.

  b. **Compatibility assessment of liquid mixes:** Before any fresh lot of liquid waste/AFR received in the plant is transferred to the liquid storage tank, it is desired to evaluate the chemical compatibility of mixing this fresh lot with the existing material in the tank. For this, the following procedure needs to be implemented.

  i. Set up a one-liter three-necked round bottom glass flask in a fume hood equipped with a stirrer, dropping funnel, and a thermometer.

  ii. Take about 300 ml of the uniform sample of the liquid contained in the storage tank and pour it in the one-liter tank.

  iii. Take a uniform sample of the fresh lot of the liquid waste/AFR received in the plant in the dropping funnel.

  iv. The proportion of the samples of fresh liquid and the tank liquid should be the same as it would be in the tank after the fresh lot is added to the tank.

  v. Add slowly, over a period of 10–15 min time, the liquid in the dropping funnel into the round bottom flask while stirring and monitor the temperature. If the temperature is seen as rising during this addition, discard the evaluation as the rise in temperature indicates incompatibility of the materials. Do not transfer the received lot into the storage tank—there could be a safety hazard.

vi. If nothing happens during feeding the liquid, continue the stirring for another 1 hour and continue monitoring the temperature. If the temperature is seen as rising during this stirring process, discard the evaluation. Do not transfer the received lot to the storage tank—there could be a safety hazard.

vii. Manage the received liquid waste/AFR separately as per the feasible option.

10.11.3 Storage of Waste/AFR Materials

Storage of AFR material consists of incoming material storage and outgoing material storage. The design of the storage facility needs to comply with the local regulation and needs to ensure the following requirement to the minimum:

A. **Protection of material from rain**

This requires a proper shed that is suitably covered with sheets from all sides and the roof: an inclined roof with a height that allows safe movement of the material handling equipment. Usually, a height of 6–7 Mts. is good enough.

A typical storage shed is depicted in Fig. 10.11.

Fig. 10.11 Storage of the waste/AFR material

Fig. 10.12 Impervious concrete flooring

B. **Protection from pollution to the soil and groundwater due to spillage**

Spillage of waste in the store area can seep into the ground causing pollution to soil and groundwater. Hence, the entire floor area of the store needs to be made of an impervious concrete floor. To protect damage to the side sheets due to operating heavy equipment, it is desired to construct concrete walls up to a height of 5 Mts.

A photo of the shed with the concrete impervious floor is depicted in Fig. 10.12.

A full proof system would be to lay the geopolymer membrane on the floor and then cast the concrete floor on the same. Further, it is required that such floor spills are transported by gravity to a storage tank from where the spill can be taken up for treatment. These drains and also the storage tank also are required to be lined up with a geopolymer membrane or at least a coating of epoxy paint to be done on all the sides to avoid soil and groundwater pollution. Further, it is required that the rainwater drains and the spill drains are ensured to remain independent of each other to avoid spills getting into the rainwater drains.

Storage of material is also carried out in bunkers and its handling is carried out using a bridge crane. Such a typical storage management facility with a bridge crane provided by Walter Materials Handling is depicted in Fig. 10.13.

Fig. 10.13 Automated bridge crane with hydraulic grab

C. Protection from spread of odour of waste/AFR

Odour, irrespective of its nature and attribute, brings huge discomfort to the operating employees and surrounding residential communities. This is required to be monitored and controlled.

Monitoring odour is to measure the concentration of volatile organic compounds (VOC) which can be done using a handheld or store-mounted device.

The most appropriate method of controlling odour from waste storage, handling, and processing facilities is to exhaust the odour causing VOCs from the concerned location into the hot zone of the kiln. When the same is not feasible, then the following are the different options which can be employed for odour control.

(i)   *Thermal or catalytic destruction of the VOCs;*
(ii)   *Microbial treatment;*
(iii)   *Fragrance spray;*
(iv)   *Adsorption of the odour causing VOCs in activated carbon or other similar adsorbents;*
(v)   *Absorption of the VOCs into suitable liquid streams;*
(vi)   *Absorption of odour causing VOCs on zeolite-coated mesh. In this system, the mesh is installed in the store and waste processing facilities in the airflow direction.*

A detailed description of the above-mentioned odour control processes is provided in Chap. 12, Sect. 12.2.6.

Figure 10.14 illustrates features of VOC absorbing mesh supplied by M/s Hiraoka that mitigate the odour issues in the waste storage handling and processing facilities

Deodoratex™ is a mesh that needs to be hung in the direction of the odor flow in waste storage and waste processing sheds. It drastically reduces the odor nuisance by absorbing various kinds of odors chemically in an effective manner. After the capacity of the mesh is exhausted, it simply needs to be washed with water and reused. It is a proven solution to deal with the odors emanating from hazardous and non-hazardous Industrial wastes and wastes from Municipal & agricultural sectors. Deodoratex™ has been installed at 26 installations worldwide and more than 10 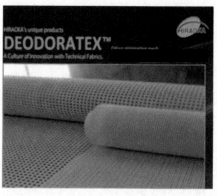 installations in India. The total area of the Deodoratex™ installed in India is 2750 Sq. M. The oldest operating installation globally is 4 years old and in India it is 2.5 years. Deodoratex™ facilitates odor free environment for the benefit of operating workforce & the surrounding community.

www.hiroaka.in   info@hiroaka.in

Fig. 10.14   Features Deodoratex Mesh for odour control

D. **Storage of incompatible materials**

Incompatible materials, if they come in contact with each other, may cause fires and explosions. Hence, they need to be stored away from each other. Since waste materials may have contaminations of different kinds and they may be received from different sources, it is better to store them material-wise and source-wise separately using barricading walls (first in first out principle helps to handle wastes safely).

The general arrangement of storage is depicted in Fig. 10.14. As shown in this figure, different materials are stored in different compartments, and these compartments are separated by concrete walls or walls made out of concrete blocks. It can be observed from Fig. 10.15 that separate entry and exit points are provided in the inbound and outbound areas.

E. **Electrical installations**

Electrical installations in the storage and processing shed must be compatible with the material being stored/handled/processed inside. Many materials would be requiring different kinds of approved electrical fittings and fixtures depending upon the flash point of the materials stored/handled. These fittings are designed to comply with the designed classification of chemicals from a hazard point of view. This classification is provided in Table 10.3. All the chemicals falling in these categories are called flammable chemicals. Those having flash points >93 °C are termed as combustible chemicals.

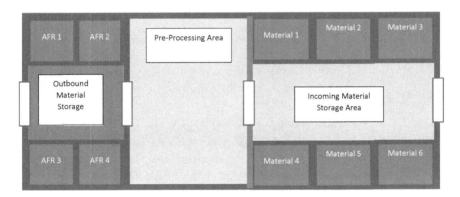

Fig. 10.15 Storage of incompatible materials

Table 10.3 Classification of Chemicals from a hazard point of view

Category	Criteria
1	Flash point < 23 °C and initial boiling point < 35 °C
2	Flash point < 23 °C and initial boiling point > 35 °C
3	Flash point > 23 °C and < 60 °C
4	Flash point > 60 °C and < 93 °C

While handling the materials falling in the above categories, specific electrical fittings and fixtures are required to be installed.

F.   **Fire Fighting Systems**

Fires are often caused at the storage sites and processing areas due to the nature of the waste/AFR materials. It is important, therefore, to set up an appropriate firefighting system to deal with such eventualities. In fighting fires, the following are the important aspects that need to be addressed appropriately in the design of the fire firefighting system.

1.   Installation of a windsock in the plant at a visible location to detect the wind direction.
2.   Installation of the fire detection and control system to deal with the fires.
3.   Installation of the alarms and hooter systems to bring awareness of the plant personnel on the emergency situation.
4.   Training of the plant personnel on the operation of the installed firefighting system.

As per IS 2190 (2010), there are four classes of fires.

a.   *Class A fires*

Fires involving solid combustible materials of organic nature such as wood, paper, rubber, and plastics where the cooling effect of water is essential for their extinction. Water, foam, ABC dry power, and halocarbons are the most suited fire extinguishing agents for this class of fires.

b.   *Class B fires*

Fires involving flammable liquids or liquefiable solids or the like where a blanketing effect is essential. Foam, dry powder, clean agent, and carbon dioxide are the most suited fire extinguishing agents for this class of fires.

c.   *Class C fires*

Fires involving flammable gases under pressure including liquefied gases, where it is necessary to inhibit the burning gas at a fast rate with inert gas, powder, or vaporizing liquid for extinguishment. Dry powder, clean agent, and carbon dioxide are the most suited fire extinguishing agents for this class of fires.

d.   *Class D fires*

Fires involving combustible metals, such as magnesium, aluminium, zinc, sodium, and potassium when the burring metals are reactive to water and water-containing agents and in certain cases carbon dioxide, halogenated hydrocarbons, and ordinary dry powders. These fires require special media and techniques to extinguish.

The fire extinguishing system could be portable or automatic. In a portable system, fire extinguishing is achieved by spraying the fire extinguishing media manually on the fire.

The automatic fire extinguishing system works on triggers or signals received from different sensory devices such as thermal expansion detectors, heat sensitive insulation, photoelectric detectors, ionization and radiation sensors, and Ultra Violet and Infrared detectors. Soon as a trigger is received to the automatic system from such sensors, the relevant fire extinguishing media is sprayed automatically on the fire.

At smaller TSR operations, the portable systems are okay to deal with the fires but once the TSR starts increasing, it is safer to implement the automatic systems. It is essential that the installed firefighting system meets the prevailing statutory requirements: A Fire Fighting system to be designed and installed as per NFPA/NBC guidelines.

10.11.4 Pre-Processing of the AFR Material

Materials that need to be used in the cement kiln as a resource must meet very stringent quality considerations to achieve smooth operation of the kiln process as well as to achieve the desired quality of the product whereas, wastes do not have quality consideration. This requires the wastes to be pre-processed to convert them into AFR materials having desired quality considerations. Pre-processing of wastes into AFRs is carried out in a properly designed pre-processing facility. Various aspects related to the pre-processing facility are illustrated in Chap. 11.

10.11.5 Feeding the AFR Material into Kiln for Co-processing

The AFR is fed in the kiln through the following four feed points as illustrated in Table 10.4.

Out of these four points, mid kiln is relevant in old technology plants only that have long wet or long dry kilns without preheater or pre-calciners. Globally, there are not many plants now that operate on this old technology. The case is similar to the separate lined calciners. They are also not very popular in modern-day cement plants. Hence, the first three points are common in modern technology dry process cement plants. All these points are located at a higher altitude and away from the waste/AFR storage location. Hence, to feed the waste/AFR materials into the kiln, separate and appropriate systems need to be built. These are different for solid AFRs and liquid AFRs.

Table 10.4 Feed points in the kiln and the suitable materials

S. No.	Feed point	Suitable materials	Examples
1	Main Burner	Liquids, fine powders, finely shredded two-dimensional materials	Waste solvents, aqueous waste liquids, spent carbon, finely shredded plastic waste, impregnated sludges with sawdust, etc.
2	Pre-calciner (In-line)	Coarse solids < 75 mm, liquids, fine powders, etc.	Shredded plastic waste, processed RDF from MSW, waste solvents, aqueous waste liquids, spent carbon, tyre chips, processed AFRs, solid AFR chunks, etc.
	Pre-calciner (Separate Line)	Liquids, fine powders, < 20 mm coarse solids	Waste solvents, aqueous waste liquids, spent carbon, shredded plastic waste, processed RDF from MSW, etc.
3	Kiln Inlet	Lumpy materials, sludges	Whole tyres, bagged materials, oil sludge, etc.
4	Mid Kiln	Lumpy materials, sludges	Whole tyres, bagged materials, oil sludge, distillation residues, greases, paint waste, soap, etc.

These are discussed in the next sections.

1.   Weighment and Transportation of AFRs to the Calciner or Main burner floor

The system required to transport Solid, Liquid, and Sludge AFR would be requiring different designs. These are described below.

A.   **Feeding of Liquid AFRs in the calciner:**
AFR pumped to the calciner floor using an appropriate pumping system is fed into the calciner using a pipe and nozzle assembly. The weighment is carried out by using a suitable flow meter or load-cell-based weight measurement system. The atomization of the liquid is achieved by pumping air into it if the same is acceptable. Else, high-pressure nozzles are made use of for atomization. It is advisable to co-process waste/AFRs having less than 1.5% Chlorine content in the calciner. from the Dioxane and Furan emission consideration.

B.   **Feeding liquid AFR in the main burner**
For liquid firing in the main burner, the modern burners already have a separate channel for firing liquid AFRs. Through this hole, a pipe having a properly designed spray nozzle is inserted into the burner. Metered quantity of liquid is pumped in the kiln using the feed-pump through this nozzle which gets atomized and burnt in the hot gases. Feeding through the main burner is mandated for waste/AFR streams having a chloride content of more than 1.5%.

C. **Solid AFRs**

There are various ways in which the waste/AFR materials are transported to the burner or calciner floor.

i. **Use of passenger lift [for properly packed waste only]**

In most of the plants, a passenger lift is available to travel to the higher floors in the cement plant. This lift travels to the burner platform floor, calciner platform floor, and other relevant floors. In the initial phases— when the waste/AFR quantum is very small, say <1%—this option may be exercised. Here, the waste/AFR material that is stored in the storage shed is manually filled in bags or containers using shovels or payloaders. The weight in the bags or containers is measured and noted.

These bags or containers are then transported using a suitable vehicle to the entry point of the passenger lift. These are then put into the passenger lift and taken up to the burner platform floor or the calciner floor as per the requirement. Later, the material in the lift is taken out and placed at a suitable location on the burner or calciner floor for co-processing.

ii. **Use of a winch-based preliminary system**

This option is suitable when the volume of waste/AFR material is small— say <2–3% TSR. In this option, weighed quantity of waste/AFR material is filled in small bins or buckets of say 0.5–2 m3 volume using manual operation or mechanical equipment. These buckets are then shifted from the storage area to the kiln area using a suitable vehicle. Subsequently, these buckets are pulled up to the main burner or calciner floor using a winch mechanism. The material in the bin is emptied on the burner or calciner floor for co-processing purposes, and the empty bin is sent down for filling the next lot. This winch-based system has lesser manual intervention than the one based on the passenger lift option.

iii. **Using a pneumatic conveying system**

This system is suitable for conveying AFRs that are fine powders such as rice husk, sawdust and other powder wastes. In this system, the waste gets transported through pipes using air as the conveying media. Compressed air is used for this purpose.

The AFR material is fed into a hopper using a suitable device which is equipped with an extraction device such as a rotary air lock (RAL) valve. The RAL feeds the material in a venture arrangement installed in a pipe line that is carrying the conveying air. The AFR material received in the venture then gets conveyed to the burner or calciner floor pneumatically.

iv. **Use of a conveyor-based standard system**

This option is suitable when the volume of waste/AFR material is large— anywhere from 3% to any higher TSR value. In this option, the waste/AFR material is conveyed to the burner or calciner floor using a long belt conveyor. The long belt conveyor is installed near the waste/AFR material storage site and on the other end reaching the burner or calciner floor.

The waste/AFR material located in the storage area is fed into a suitable material extraction system such as a walking floor and extractor. Then this material from this extraction devise gets weighed in a weigh feeder and then is dropped onto the long belt conveyor using different systems such as bridge/grab crane, chain belt conveyor, and short belt conveyor. The long belt conveyor transports the waste/AFR volume to the burner/calciner floor and drops onto another belt conveyor which is a small one and is called a sacrificial belt conveyor. Usually, the angle of the long belt conveyor is maintained at ~13°.

A.   **Liquid AFRs**

The most common scheme for liquid waste/AFR is the use of liquid pumps. The liquid waste/AFR materials that are stored in tanks located in the storage area are pumped at a defined rate using different types of pumps. These pump types depend upon the liquid characteristics. The pumps that are used for pumping are diaphragm pumps, slurry pumps, centrifugal pumps, gear pumps, etc.

For achieving this pumping, piping—with appropriate valves—is laid out from the storage tank to the burner entry point. Depending upon the situation, a coarse and fine filtration arrangement is also implemented to facilitate smooth pumping operation. The material of construction of the pipes, valves, fittings, and tanks depends upon the corrosion characteristics of the waste/AFR material to be handled in them. The MOC can be of different kinds such as carbon steel, stainless steel, rubber lined carbon steel, teflon lines carbon steel, PVDF lined carbon steel, and different kinds of plastics.

B.   **Sludge AFRs**

Sludge AFRs are very viscous in nature. They are usually impregnated with suitable materials and then transported to the calciner floor using the schemes mentioned in section A above.

It is also possible to pump these sludge AFRs directly to the calciner floor using a pump which is similar in construction to the one used for pumping the concrete mix. The material piping is suitably routed to the calciner feed point with a minimum number of bends to reduce the pressure drop.

III.   **Feeding arrangement of AFRs in Calciner, main-burner, and Kiln Inlet**

IV.   **Feeding Solid AFRs in calciner and kiln inlet**

The solid material conveyed to the calciner floor is then fed onto a short belt conveyor named as a sacrificial belt. The material from the sacrificial belt is then dropped into a double flap valve or a rotary valve or screw conveyor-based feeding device which is coupled with a shut gate. The feeding device and shut gate assembly are mounted on the chute leading into the calciner. Figure 10.16 provides the views of the double flap valve, rotary valve, screw conveyor, and shut-off gate.

i.   **Double Flap Valve with a shut gate**

This system consists of a hopper below which a double flap valve (DFV) is attached. These two flap valves open and close alternatively thereby

Fig. 10.16 Feeding devices
for solid AFRs

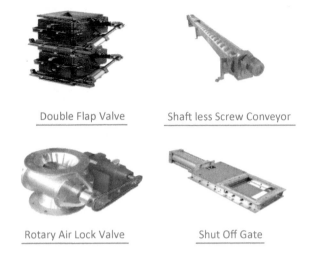

Double Flap Valve   Shaft less Screw Conveyor

Rotary Air Lock Valve   Shut Off Gate

retaining material for a while before dropping it down. Below the DFV is a shut gate that normally remains open. It gets closed wherever the temperature or the pressure in the nearby duct area increases above a defined setpoint value. The material is first dropped on a sacrificial belt conveyor. This belt conveyor is called a sacrificial belt conveyor because this small conveyor only is foreseen to catch fire and restrict the damage to a small level.

V. **Rotary Airlock system**

The rotor assembly in a rotary airlock valve consists of a set of blades welded to a rotary shaft. A six-vane rotor has six metal blades attached to the shaft; an eight-vane rotor has eight. A valve with more vanes tends to have greater sealing ability than one with less ones. The material fed in the vane at one end gets moved with the rotor and when it reaches the other end, it gets dropped.

VI. **Screw conveyor-based system**

In this system, a screw conveyor is utilized to feed the material from a hopper into the calciner. This screw is normally a shaftless screw. In this arrangement, the material from the sacrificial belt drops in the hopper of the screw conveyor from where the rotating screw below the hopper conveys the material inside the calciner.

VII. **Shut Gate system**

The shut gate system is provided to cut off the AFR feed in case of an emergency situation which causes hot gases present in the kiln with high temperature to come out of the kiln system through the chute and cause fires. These emergency situations would be positive pressure in the kiln system or high temperature in the feed chute. This shut gate system is usually a pneumatically operated device. It consists of a plate that moves in a slot mounted in the chute. This plate slides in the sliding slot due to the compressed air system. In an emergency situation, the plate of the shut-off gate slides in the slot due to the air pressure and closes. This makes the feeding system to get completely closed.

Various commercial systems are available for achieving proper feeding of the materials in the calciner. These are depicted in the next pages as in Figs. 10.17, 10.18, and 10.19, respectively.

Figure 10.17 depicts the Doseahorse by Walter Materials Handling that is typical equipment which helps to feed the solid material at a constant rate.

A double valve with a shut gate supplied by Walter Materials Handling is depicted in Fig. 10.18 along with intricate details of the double valve for a better understanding of the system.

Also, in Fig. 10.19 the picture and description of shaftless screw-based material feeding system called Multiplex of Schenck Process is depicted for its better understanding.

VIII. **Feeding Sludge AFRs in the kiln inlet:**
Generally, sludges are pumped in the kiln inlet or in the riser duct and a concrete type pump is used for pumping purposes. The sludge is fed into the hopper of the pump and the pump pushes the sludge through the pipe and drops it into the kiln inlet or riser duct in which the feed pipe is inserted.

IX. **Feeding lumpy AFR in the kiln Inlet**
To facilitate the feeding of lumpy AFR into the kiln inlet, a double flap valve and shut gate arrangement are implemented. The whole tyres or bagged materials are then dropped into the kiln inlet through this double flap valve and shut gate arrangement. The lumpy material is conveyed to the kiln inlet floor using a winch arrangement or can be done through a belt conveyor.

X. **Advancements in Pyro-systems for AFR co-processing**
The modern burners have an additional channel for firing solid AFRs. Solid AFRs are fed through this channel using air as the media. A typical design of the modern-day burner by KHD PYROJET® is provided in Fig. 10.20. The most imperative feature of such a burner is its short, intense, and stable flame.

For co-processing difficult to combust materials with given limitations in the existing kiln systems and also to co-process less pre-processed or un-processed waste streams, new advanced pyro-systems are being designed by various system suppliers. One such system is PYROROTOR®. With a PYROROTOR®, an unmatched level of flexibility for alternative fuels paired with a significant potential for cost reduction in combination with very low or even un-processed secondary fuel with a maximum shift of calciner fuel to this special design state-of-the-art rotary processor is reaped. This is depicted as in Fig. 10.21.

Another feature is the KHD PYROCLON® Calciners with Combustion Chamber. This comes as an integral part of KHD's long duct type Pyroclon R calciner which is well proven to process the coarser green fuels due to its design feature of higher gas velocities inside the calciner making the system more selective than other equivalent

Fig. 10.17 Doseahorse for feeding AFR uniformly

technologies available in the cement industry. This is depicted in Figs. 10.22 and 10.23.

The key features of this burner with AFR Swirl Nozzle for Alternate Fuels and NOx reduction are retractable swirl element, no additional primary air necessary for AFR nozzle, adjustable during operation, fracturing and mixing of the AF flow

ELECTRICAL DOUBLE VALVE AIRLOCK WITH SAFETY SHUT OFF GATE

For safe, reliable, flexible feeding of alternative fuels into calciner

The Introduction of alternative fuels into the cement manufacturing process is defined as co-processing. This introduction is challenged by many factors of the cement manufacturing process. One is extreme high temperature of the calciner about 1100 degree Celsius and the second challenge is to seal the entry of ambient air entry into the calciner, which is a loss of energy for the system. To solve the issue, Walter materials of ATS-Group designed a unique mechanism called double valve with electrical drives.

Electrical double valve airlock with pneumatic safety shut-off gate designed by Walter materials of ATS-Group, emerged as the best solution for feeding alternative fuels safely & efficiently. These double valves now more than two decades working in many cement plants globally. Happy end-users with many repeated customers are the clear sign of the success of this equipment for the said process application.

Arrangement of two valves with safety gate

The opening closing of valves is realized by pendulum of a single flap. An electrical motor ensures a very short cycle time, efficient closing and opening. VFD regulates the acceleration and deceleration of the valves, so that flaps open and close smoothly and rapidly. Sensors are used to indicate each position of flap. Technical advantage of electrical drive managed with quick operation over pneumatic actuator is more than 50% reduction in cycle time. Such a double valve system enables safe introduction of alternate solid fuel of varied nature (Grain size 1 to 400 mm) into Calciner with short cycle time and minimizing introduction of false air.

Regarding security, the double valve airlock offers three safety levels. Level 1 and 2 are in case of power shut down, the mechanical flap closing is ensured with integrated counterweight. Level 3 – in case of pressure drop from plant air tank placed on electro pneumatic cabinet ensured that safety gate is closed. Excessively high temperature will be detected by sensors located between top and bottom valves. Mechanical performance of double valve airlock is also due to right choice of material. This new concept of double valve airlock is very compact and can be mounted easily on the Calciner floor.

Single valve internal construction

Please visit our web for more & latest details of our offerings: www.ats-group.com

Reach us at: contact@ats-group.com, wm@ats-group.com, ats.india@ats-group.com

WALTER Materials handling

ATS Group

Fig. 10.18 Electrical double valve airlock with safety shut-off gate

MultiFlex NG - Flexible & Efficient Handling of Alternative Fuels for Cement

Cement producers face the challenge of how to lower their carbon footprint economically, as well as improve their contributions towards sustainability by adopting alternative fuels. However, to do so – they ideally need to be capable of productively and accurately handling mixed fuels. Helping to address this issue, Schenck Process have engineered a dynamic feeder especially for the cement industry – the MultiFlex NG Feeder. Alternative fuel materials used in the production of cement, can come in a wide variety of thicknesses, shapes, coarseness and types. Materials may vary from household waste, dried sewage sludge and derivative fuels,

through to organic waste and tyre chips. Having a feeder that is capable of flexibly adapting to process materials with these variabilities, without causing unnecessary maintenance and downtime, is difficult. Schenck Process have therefore designed the MultiFlex NG Feeder with a number of features that makes processing mixed fuels - efficient, precise, consistent and cost effective.

Initially to control the filling, an advanced weighing system allows automatic calibration. Continuous calibration of the combined weighing system has been engineered into the MultiFlex Feeder, to yield long term accuracy of +/- 1%. This leads to less system wear from varying materials, longer life performance and lower energy consumption. To ensure accurate and proper filling of the screws, the material inlet of the screw trough is weighed; also enabling control of speed. Material bridges in the hopper are detected and activating agitators maintain correct feeding of the trough. This occurs only when necessary, for high energy efficiency and system lifespan. The screw trough is weighed prior to material discharge, thereby controlling the mass flow of material, relative to the speed of the conveying screws. Synchronising sensors continuously adjust the phase of each conveying screw, to prevent material bundling when feeding variable materials and to maintain a smooth flow. For small feedrates, single or double screws, allow for a wide control range.

Other features of the MultiFlex NG include ATEX and reduced explosion design, as well as robust construction for maximum wear protection. Ultimately, cement producers are under more pressure than ever to ecologically improve their production processes. Using a variety of alternative fuels to do so, demands both efficiency and stability. The flexible design of the MultiFlex NG Feeder from Schenck Process, not only allows for high accuracy and consistent feeding of many materials, but also increases the operational lifespan of the system. Supporting the ethos of enabling a greener future, the MultiFlex since its launch of March 2021, has turned approximately 2.6 million tonnes of plastic into energy. While, every hour it helps save the environment from 500 tonnes of plastic. All these are ecological benefits that cement producers can also reap from and pass on.

www.schenckprocess.com

Fig. 10.19 Description of multiplex

Fig. 10.20 KHD PYROJET
BURNER

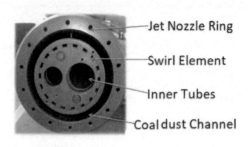

Jet Nozzle Ring

Swirl Element

Inner Tubes

Coal dust Channel

Fig. 10.21 KHD
PYROJET®

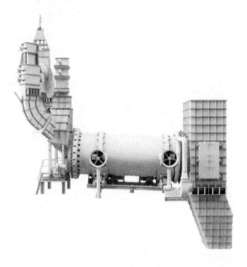

shortly before entering the kiln (increasing the spraying angle), and improving the
mixture within the flame core and with required combustion air is the key design
highlight which makes the equipment suitable for processing of secondary fuel in
the rotary kiln.

KHD PYROROTOR®is a unique rotary reactor that uses tertiary air as combustion air to process materials with inferior burning properties reliably as a secondary
fuel for cement production. Due to directly coupled direct drive, the combustion
reactor enables constant material movement. Combined with adjustable rpm and
long residence times, the PYROROTOR® enables complete burn-out of alternative
fuels such as low-processed RDF, coarsest waste matter, and even whole tyres.

KHD PYROCLON®**Calciners with Combustion Chamber** are apt for utilization of coarse fuels with poor ignition and burning properties like coarse anthracite,
petcoke, and coarse secondary fuels or waste-derived fuels. The characteristic
features of the combustion chamber are the ignition and start of combustion in pure
air at high temperature (T > 1200 °C) maintaining the calciner retention time > 7
sec. Apart from high efficiency and flexibility, it offers lower demand on fuel quality
and preparation efforts.

Fig. 10.22 KHD PYROCLON® calciners with combustion

10.11.6 *Flow Schemes for Co-processing of AFRs with Different Feeding Arrangements*

Depending upon the nature of AFR material, the capacity of co-processing desired, the space available to implement the system, and the budget available to implement the system, there can be different flow schemes with which the co-processing can be implemented. The following four flow schemes illustrate this concept for clarity.

Flow sheets of the following co-processing schemes are provided and discussed below for clarity:

a. Flow scheme of co-processing Solid AFRs using a winch-based preliminary system.
b. Flow scheme of co-processing Solid AFRs using conveyor-based standard system.
c. Flow scheme of liquid AFR firing system into the main burner.
d. Flow scheme of Solid powder and shredded plastics firing in the main burner.

a. Flow scheme of co-processing Solid AFRs using winch-based preliminary system

Fig. 10.23 KHD
PYROJET® calciners with
combustion

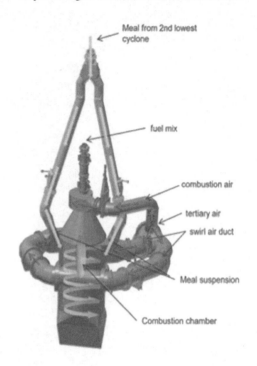

In this scheme, the solid AFR material received in the store is filled in the bucket attached to the winch and is pulled up using a motor system. This material is then dropped is a hopper that is mounted above the double flap valve. From the hopper, the material gets fed into the calciner or kiln inlet through a chute. A shut gate is installed between the chute and the flap valve to prevent the hot gases from coming out of the system in case the pressure in the kiln system tends to become positive. The flow sheet of this scheme is depicted in Fig. 10.24.

b.   Flow scheme of co-processing solid AFRs using conveyor-based standard system

In this scheme, the solid AFR material received in the store is conveyed to the preheater floor using a conveyor. To achieve uniform flow feed, a walking floor and a weigh feeder are used. The conveyed material is then fed into the kiln through the flap valve and shut gate assembly. To protect the long conveyor from possible fires, a sacrificial conveyor is used. This flow scheme is depicted in Fig. 10.25.

iii.   Flow scheme of liquid AFR firing system into the main burner

In this scheme, the AFR material received in the store is conveyed to the preheater floor using a conveyor. To achieve uniform flow feed, a walking floor and a weigh feeder are used. The conveyed material is then fed into the kiln through the flap valve and shut gate assembly. To protect the long conveyor from possible fires, a sacrificial conveyor is used. This flow scheme is depicted in Fig. 10.26.

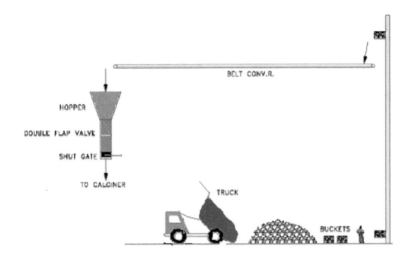

Fig. 10.24 Typical flow sheet of a winch-based preliminary co-processing system

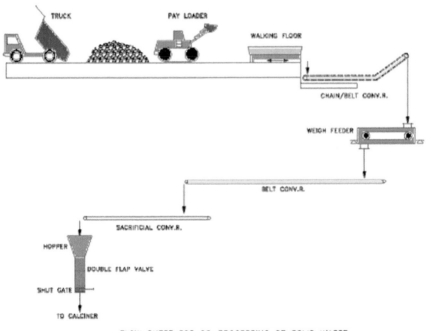

Fig. 10.25 Typical flow sheet of a conveyor-based standard system

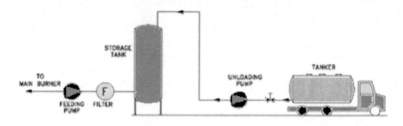

FLOW SHEET FOR CO-PROCESSING OF LIQUID WASTE IN MAIN BURNER

Fig. 10.26 Typical flow sheet of a liquid firing in the main burner

iii.   Flow scheme of fine solid AFR and shredded plastics firing in the main burner

In this scheme, the fine solid AFR and the shredded plastic material of <10 mm size
are received in the plant and conveyed to the burner platform using a winch system.
Subsequently, it is fed in a hopper, from where the rotary airlock drops the material
into a venturi device. This material is then blown into the main burner using air.
    This flow scheme is depicted in Fig. 10.27.
    The systems set up for the co-processing of AFRs need to be flexible and should
be able to handle and manage a variety of AFRs with the same system. A case study
of the Multiplex equipment of Schenck Process is described in Fig. 10.28.

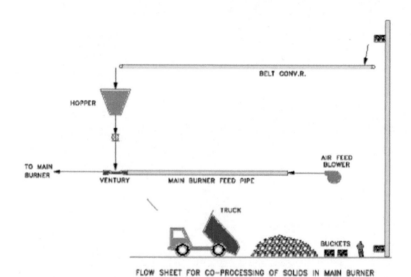

FLOW SHEET FOR CO-PROCESSING OF SOLIDS IN MAIN BURNER

Fig. 10.27 Typical flow sheet of a solid firing in the main burner

Case Study : Process Enable Turnkey Systems in Alternative Fuels

To ecologically enhance the processes of cement production for a key client, Plzenska Teplarenska, Schenck Process integrated and customized turnkey projects for two alternative fuel feeding lines in a heat and power plant.

Customised for the plant; the line feeding wooden pellets for boilers, contains among others, fuel reception units, for walking floor trailers which also serve as operation storage in a just-in-time system. From the reception, the fuel is transported through screening of oversize lumps to the boiler house by a tube conveyor. In the boiler house, an internal conveying and distribution system is installed, transporting the fuel to weighed intermediate storage bins - distributing the fuel to the mills of each boiler. Alternatively, the fuel may be fed directly to the burners through the supplied pneumatic feeding systems.

A line for feeding coal sludge contains reception stations suitable for the rotation discharging of standard containers, storage silo and tube conveyors transporting the material to the boiler house. An internal transport and feeding system of the fuel to the boiler is also installed. Concept of this line allows universal utilization and a large variety of fuels to be fed through this line.

www.schenckprocess.com

Fig. 10.28   Case study on universal use of multiplex for different alternative fuels

10.12   Other Relevant Considerations in Co-processing

Co-processing attracts three specific considerations. The first one concern the regulatory requirements, the second one concerns the liability associated with the waste materials, and the third one corresponds to the sustainability considerations.

10.12.1   Statutory Considerations

Wastes are legally controlled materials and they are governed by rules and regulations. It is important to comply with them while undertaking activities related to their transportation, handling, processing, etc. The wastes that are feasible to be co-processed belong to Municipal, Agricultural, and Industrial sectors. The regulatory provisions pertaining to them are clearly defined in Hazardous and Other Waste Management Rules, Plastic Waste Management Rules, and Solid Waste Management Rules that are amended by the government from time to time. They are also governed by various guidelines published by the Government in respect of the relevant wastes from time to time.

10.12.2 Liability Considerations

Wastes carry liability with them and when wastes get transferred from one stake-holder to another, the liability associated with them also gets transferred from one to another. Wastes cause substantial damage to the environment; they also bring about huge social and economical concerns due to their odour, toxicity, flamma-bility, corrosivity, etc. It becomes important therefore that while undertaking co-processing, these concerns are appropriately addressed. All precautions need to be taken to ensure that environmental damage, societal impact, and financial losses are avoided while managing wastes through co-processing. As the wastes have varying physico-chemical characteristics, it is difficult to prepare their MSDS. However, it is important to evaluate and compile the relevant Safety, Health, and Environment-related parameters so as to ensure safety in co-processing. This requires the following approach:

- Obtain the SOPs implemented in storage, handling, and processing of wastes from the waste generators.
- Study the safe practices implemented by the waste generator at his end and compile appropriate practices.
- Compile the literature information related to the health and safety aspects of wastes.
- Discuss with the factory doctor associated with the waste generator on aspects related to health impact.

10.12.3 Sustainability Considerations

Sustainable management of wastes is an important consideration from various considerations such as resource efficiency, environmental degradation, conservation of natural resources, and mitigating climate change impacts. For managing wastes therefore waste management hierarchy is the guiding principle. Figure 10.29 depicts the waste management hierarchy that needs to be considered for sustainabile waste management.

In the waste management hierarchy, landfilling and incineration are the last options to be pursued when all other options have failed to get implemented. Waste to Energy is lower in the hierarchy because the material recovery is 0% and energy recovery is < 25%.

It is important to note that recyclable materials and cattle feed materials are not co-processed because recycling and cattle feed are higher in waste management hierarchy than co-processing.

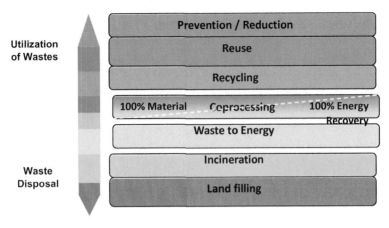

Fig. 10.29 Waste Management Hierarchy

10.13 Challenges Faced in Co-processing of AFRs

Compared to the conventional input materials, AFRs have substantially different characteristics and hence many challenges are faced while co-processing them.

10.13.1 Specifications and Availability of the AFR Materials

The specifications of the conventional materials can be defined for ordering and material can be received as per the same. Further, the properties of the material are generally uniform from lot to lot. The number of input streams of conventional materials required for cement manufacture are generally less than ten. Further, they are available in large quantities as desired for cement manufacture. However, AFRs would have to be accepted as per their available specifications; there is substantial variation in their physico-chemical characteristics from lot to lot and generally, they are available in small quantities only. To achieve the desired quantum requirement of AFR in cement manufacture, a large number of waste streams—running into hundreds—need to be utilized.

10.13.2 Increase in the Specific Thermal Energy Consumption

There are several reasons for the increase in the specific thermal energy consumption while co-processing AFRs. These include the following:

a.   AFRs have higher ash and moisture content.
b.   Their burnability is poor compared to conventional materials which requires higher oxygen level to burn.
c.   Fluctuating feed rates and variation in the AFR quality.
d.   Circulation of some of the constituents in the kiln system.
e.   False air ingress into the kiln through the new feeding ports.

10.13.3   Impact on the Emissions from the Kiln

Generally, the emissions from the cement kiln do not get influenced due to the co-processing of AFRs if the general principle of monitoring inputs is implemented properly. The impact on the emissions would be noticed when volatile heavy metals are not appropriately controlled in the AFR and also if AFR containing volatiles is fed through the raw material feed route.

10.13.4   Coating in the Preheater Section of the Kiln

Chlorine, alkali, and sulphur content in the AFR influence the coating tendency in the preheater section, and this needs to be properly monitored in the hot meal and controlled.

10.14   Conclusion

Co-processing is a very powerful technology tool to reduce the environmental pollution load, conserve natural resources, and reduce the GHG emissions. It needs to be carried out in a manner that ensures monitoring and control of inputs, processes, products, and emissions. Co-processing can be implemented for hazardous and non-hazardous industrial wastes, POPs, agricultural wastes, and wastes derived out of municipal activities. Pre-processing of these wastes as AFR is an essential requirement for successful and gainful co-processing without making major impacts on the process parameters. There are appropriate feed points in the kiln which need to be selected depending upon the nature of the AFR material. Occupational and Environmental Safety is of prime consideration in the co-processing initiative. Co-processing of AFRs may cause an impact on the kiln operation and emissions and to avoid or reduce the same, monitoring and control on the AFRs needs to be put in place. AFR co-processing also causes an increase in the specific thermal energy consumption of the kiln and necessary measures need to be put in place to keep it to a minimum.

References

Guidelines on Co-processing of Waste Materials in Cement Production (2006) Published by GTZ & Holcim Available at https://www.geocycle.com/sites/geocycle/files/atoms/files/co-processing_supporting_document_giz-holcim_guidelines_0.pdf

Technical guidelines on the environmentally sound co-processing of hazardous wastes in cement kilns published by Basel Convention in 2011.

Updated general technical guidelines for the environmentally sound management of wastes consisting of, containing or contaminated with persistent organic pollutants (POPs) published by Basel Convention.

Guidelines for Pre-processing and Co-processing of hazardous and other wastes in cement plant as per H&OW (M & TBM) rules, 2016 published by CPCB, GOI, July 2017.

Ulhas Parlikar et al. (2016). Effect of variation in the chemical constituents of wastes on the co-processing performance of the cement kilns. *Procedia Environmental Sciences,35,* 506–512. Retrieved from https://www.sciencedirect.com/science/article/pii/S1878029616301244

Chapter 11
Pre-processing of Wastes into AFRs

11.1 Introduction

Pre-processing of wastes into AFRs is an important consideration in achieving successful co-processing of AFRs. As a broad categorization, wastes are solids, liquids, or sludges. They can also be in gaseous forms but the cases are very few. Wastes, by their very definition, do not have any quality consideration. They generally have very large variations in physical and chemical characteristics. For example, municipal solid waste (MSW) from one household is very different than that from another household. The same from the same house will be varying on a day-to-day basis. If dry waste is segregated from the MSW, then its composition also is different from house to house and time to time. If the recyclable materials are separated from the dry waste, the remaining non-recyclable material is a very good fuel material having calorific value. This material due to heterogeneous nature will be having different size, calorific value, ash content, moisture content, chloride content, etc., on a lot-to-lot basis. If this waste is fed into the cement kiln on an as-received basis, for use as Alternative Fuel, it causes severe process disturbances as well as impacts the quality of clinker being produced from the kiln. Hence, prior to feeding the waste materials in the kiln, it needs to be processed to achieve uniform physical and chemical characteristics. This processing of waste materials having varying quality considerations into AFRs having uniform quality consideration is called pre-processing.

11.2 Permitting

Pre-processing involves the receipt, storage, testing, handling, and processing of wastes physically. The wastes include both hazardous and non-hazardous ones. Although pre-processing does not involve any thermal treatment or chemical reactions, there are possibilities of the release of effluents and emissions and as per the

© The Author(s), under exclusive license to Springer Nature Singapore Pte Ltd. 2022   243
S. K. Ghosh et al., *Sustainable Management of Wastes Through Co-processing*,
https://doi.org/10.1007/978-981-16-6073-3_11

local regulations, permission for implementing the pre-processing facility needs to be taken and the prescribed norms need to be complied with. In most of the countries, specific waste-wise permissions may also be required in respect of receiving, storing, and pre-processing them.

11.3 Unit Operations in Pre-processing

The pre-processing involves different unit operations such as blending, shredding, drying, impregnation, size separation, segregation, and bailing. In pre-processing, mostly mechanical treatment is carried out. However, sometimes chemical treatment processes such as neutralization, flocculation, and sedimentation are also implemented.

11.3.1 Size Reduction or Shredding

Size reduction of different materials is achieved by using shredders. There are three types of actions that facilitate shredding, namely Shearing, Tearing, and Fracturing. Shearing involves the cutting of material with scissor-like tooling, tearing involves pulling the material with great force, and fracturing involves breaking of the brittle materials using impact. Some examples of materials that are commonly shredded are tyres, metal cans and sheets, car wrecks, wood, plastics, leathers, papers, clothes, industrial hazardous and non-hazardous wastes, municipal solid waste, refuse derived fuel, etc.

An industrial shredder is a piece of heavy-duty equipment designed to shred dense and light materials to prepare them for gainful utilization. It is a device that transforms a waste material having non-uniform size configuration into a resource having defined size configuration. A shredding device is used for materials such as metals, plastics, aluminium, metal and cars, and wastes from municipal, nuclear, and medical sectors. These waste materials can be hazardous and non-hazardous.

Figure 11.1 depicts the picture of a typical industrial shredder.

The shredding process consists of feeding the material into the hopper of the shredder from the top using a feeding device or a conveyor. The fed material gets grabbed by the blades and pulled in through their rotating action and causing shredding operation by the blades. Many of the shredders also have a built-in screen to allow only the material smaller than the size of the screen to get released from the shredder and the higher size is sent back to the blades. The shredded material subsequently gets released from the shredder through different mechanisms such as gravity and pneumatics.

Most of the industrial shredders are capable to handle metals that are present in the feed material. These metals have good recycling value and hence are segregated from the shredded mass using magnetic and eddy current separators. The feed material

Fig. 11.1 Typical industrial shredder

would also be containing inert material such as stones and grit which is having higher specific gravity than the combustible material. The separation of this inert material from the combustible material is achieved using pneumatic separation techniques. Industrial shredders that do not have a screen built inside. It delivers shredded material that does not have specific size configuration. If the size of the shredded material is a required criterion, then the output from such shredders is sent to screening systems that are mounted external to the shredder. The shredded material that passes through the external screening equipment is collected as the desired material and the oversized material from this screening device is fed back to the shredder through the hopper using material handling equipment or conveyor system.

Industrial shredders come in many different design variations, sizes, and applications. The industrial shredder has many components. These are rotor, counter blades, housing, motor, transmission system, power system, and electrical control system. They can be a single shaft, twin shaft, or multi-shaft and may be driven by electric power or hydraulic power. The blades of the shredder are of different designs and materials to cater to the varying hardness of different waste materials. These shredders are also classified as metal shredders, plastic shredders, tyre shredders, scrap shredders, aluminium shredders, wood shredders, etc.

Single Shaft Design   Double Shaft Design

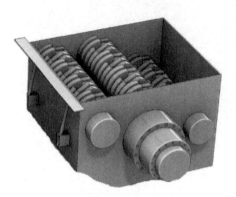

Three Shaft Design   Four Shaft Design

Fig. 11.2   Different types of shredder designs

An industrial shredder can be equipped with different types of designs. Figure 11.2 depicts different types of shredder designs available for shredding purposes.

A.   **Single-Shaft Design**

These shredders have one shaft with rotary blades, a hydraulic pusher plate, and a screen underneath to filter materials to conform to the proper size. The single shaft rotates at a low speed and shreds materials to one or two inches. They are used when a consistent particle size is required and are ideal for shredding plastic materials.

B.   **Double-Shaft Design**

The double shaft shredders have shearing blades mounted on two shafts that rotate into each other at slow speeds to quietly shred large high-volume feedstock into small pieces of one–five inches. The low speed prevents the creation of dust during shredder operation. The main purpose of double-shaft shredders is to handle large quantities of bulk volumes of materials.

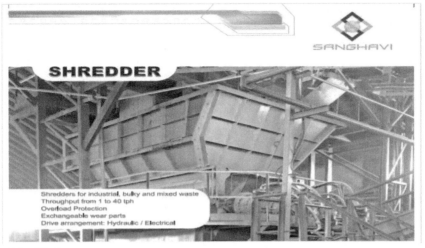

Salient Features

(1) Heavy Duty & Robust Design; (2) Replaceable Cutting Knives; (3) Output fraction from 75 mm to 100 mm; (4) Low speed & High Torque design; (5) Economical Solution for Shredding Industrial & Bulky waste

www.sanghavigroup.com

Fig. 11.3 Industrial double-shaft shredder by Sanghavi Engineering Pvt. Ltd.

Description of a commercial double shaft shredder manufactured by M/s. Sanghavi Engineering Pvt. Limited, India, is depicted in Fig 11.3.

C.   Three-Shaft Design

In the three-shaft design shredder, blades of three shafts rotate at different speeds to provide a continuous flow of feedstock. The size selection for shredded materials is determined by the screen that the material has to pass through when leaving the shredding chamber. If the material is not small enough, it is recirculated through the machine until it is the proper size to pass through the screen.

D.   Four-Shaft Design

A four-shaft shredder has four shearing rollers with four sets of shearing knife rollers with different cutting shapes. The process of a four-shaft shredder allows for pre-shredding and secondary shredding to happen simultaneously, which improves production efficiency. Quad- or four-shaft shredders are used to shred materials that need separation with uniform-sized particles.

Several manufacturers supply the shredding systems having different designs as above.

Fig. 11.4 Typical Hammer
Mill Design

E.   Horizontal and Vertical Hammer mill

A hammer mill has a drum in which hammers are mounted. These hammers are fixed
to the central rotor and are free to swing. The rotor spins at high speed and reduces
the size of the fed material due to impact. The material is then expelled through
screens of a select size that is mounted on the drum.

Figure 11.4 shows a typical hammer mill in action.

F.   Grinder

A grinding machine or grinder is an industrial power tool that uses an abrasive wheel
for cutting or removing the material. Two styles of grinders are tub and horizontal.
Tub grinders are top-loading and are designed to grind wide materials. Horizontal
grinders have a conveyor belt and do a smooth consistent grinding. Grinder's shave,
chip, and grind small pieces from large objects using abrasives or compression that
flattens the material. Grinders break materials down to fragments that are a half-inch
or less.

G.   Granulator

Granulators have an electric motor that turns a rotor having cutting blades attached
and enclosed in a chamber. They come in a wide variety of sizes and shapes. In the
chamber, the blades on the rotor shred the material and turn it into flakes or granules.

H. Other considerations in waste treatment or waste conditioning systems

The design of the waste treatment or waste conditioning systems needs to consider several different aspects such as the following:

- Ingress of other types of materials such as metals, glass, and concrete in the waste stream.
- The waste materials may have hazardous characteristics. They may be corrosive, flammable, reactive, toxic, etc. The waste may have low flash point mearing materials as well.
- The equipment needs to have proper consideration towards the ATEX requirements.
- Size reduction requirements for coarse and fine sizes. Shredding of plastic waste or RDF to less than 50 mm size would be feasible with a slow-speed primary shredder, whereas shredding the same to less than 30 mm may require high-speed secondary shredder.
- Dust containment and removal system to avoid environmental pollution.
- Desired size fraction requires screening of the shredded mass and recycling the oversize. This screening arrangement may be present in the shredder itself or may require to be externally installed depending upon the design of the shredder.
- The portability of the shredder also could be an important consideration depending upon the requirement. These can be truck-mounted as well.

The design of the treatment system needs to take into consideration the above-mentioned aspects to have a smooth, trouble-free, and environmentally sound operation (www.wasteconditioning.com).

Figure 11.5 provides a description of such a waste conditioning system designed and supplied by Loesche GMBH. This system caters to the alternative fuel needs of the cement and power industry as well as it facilitates the manufacture of high-quality pellets for use in gasification plants producing diesel, methanol, or hydrogen.

11.3.2 Drying

Industrial dryers help in removing moisture present in different kinds of materials on a large scale. Industrial dryers come in many different models depending upon the type and quantity of material to be processed. The selection of the appropriate type of dryer depends upon the factors such as the nature of the material to be dried, final product quality requirements, volume of material to be dried, and available space for installation.

The dryers are utilized to dry all three types of waste materials, viz., liquids, solids, and sludges. The material to be dried includes aqueous waste streams, organic liquids contaminated with moisture, plastics, combustible waste materials, industrial hazardous and non-hazardous waste streams contaminated with moisture, municipal solid waste, etc.

As an established expert in material conditioning for the heavy industry, Loesche saw the need to extend its products and services further back in the supply chain, which naturally led to the topic of alternative fuels. By understanding the end process very well, Loesche is able to tailor-make fuels which suit the specific needs of each process. Such processes range from alternative fuel utilization in cement and power plants, up until complete waste conditioning plants producing high quality pellets for gasification plants producing diesel, methanol or hydrogen. The picture below shows such a process – beginning with the receiving bunkers (1), the waste is transported to pre-shredders using front loaders. From then, off-spec material such as large pieces and textiles are manually removed (2). On the next step, all other unwanted impurities are removed, and recyclable material is sorted out of the main stream by using screening, wind sifting and optical sorting steps (3). After that, the material is ground in A TEC Rocket Mills®, which creates a high-quality alternative fuel (4). Due to its characteristics of ripping the material apart instead of simply cutting it, the product shows a higher specific surface when compared to traditionally conditioned alternative fuels, which brings numerous advantages when increasing substitution rates. Separation of foreign materials and partial drying also occur within the Rocket Mill, further increasing product quality. After leaving the Rocket Mills, the material can be further dried if required, and then either loaded onto trucks in bulk, or densified using balers or pelletizers (5). The off-spec material is conveyed to a reject bunker, where it will be transported to a sanitary landfill.

Loesche GmbH
Please visit us at www.wasteconditioning.com
E-Mail: waste.conditioning@loesche.de
Phone: +49 211 -5353-0
Fax: +49 211-5353-500

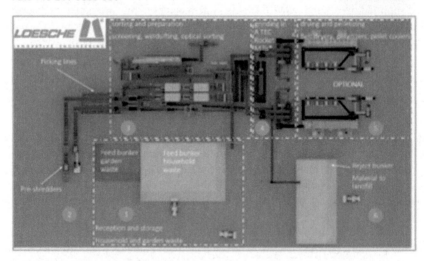

Fig. 11.5   Waste Conditioning System offered by Loesche GMBH (www.loesche.com)

Drying requires various sources of energy. The thermal energy required for drying is obtained from waste heat available in the plant or is generated using different fuel sources. These fuel sources include fossil fuels such as coal, oil or gas, biomass other alternative fuels, and solar energy.

The common types of industrial dryers utilized for the drying of various kinds of liquid, solid, and sludges are fluidized bed dryers, rotary dryers, rolling bed dryers, spray dryers, conduction dryers, convection dryers, etc. These are briefly described below.

Figure 11.6 illustrates different kinds of dryers utilized for drying different kinds of AFRs.

A. *Fluidized Bed Dryers*

A fluid bed dryer works on the principle of fluidization. In the fluidization drying process, hot air or hot gas flow is flown through a perforated plate and introduced through the bed of solid particulates that are supported onto the perforated plate. This hot gas or hot air will move upwards through the spaces between the particles and as

Fluidized Bed Dryer

Rotary Dryer

Rolling Bed Dryer

Spray Dryer

Fig. 11.6 Different types of dryers utilized for drying AFRs

the velocity of the gas is increased, the material gets fluidized after the drag forces on the particles increase beyond the gravitational forces acting on the particles. The particles at this gas-suspended stage get dried due to contact of hot gas with the wet material.

The fluidized bed dryer can be a batch type or continuous type. Different waste materials that can be dried in this type of dryer are industrial hazardous and non-hazardous solids chunks and powders, wood and biomass, plastics, RDF, MSW, etc.

B. *Rotary Dryer*

The rotary dryer works on the principle of heating the material that is rotating inside the drum due to the rotation of the drum. The heating media utilized is hot gas or hot air. This thermal energy utilized in the dryer can be waste heat or hot gas or air generated using fossil fuels, alternative fuels, or regenerative heat sources such as solar.

The rotary dryer is made up of a rotating drum, a drive mechanism, and a support structure. The cylinder is inclined from the feed end to the discharge end. The material to be dried enters the dryer and as the dryer rotates, is lifted by a series of flights mounted on the inner wall of the dryer, and falls down. While the material is falling, it comes in contact with the hot air and gets dried. Various materials that can be dried using this dryer are industrial hazardous and non-hazardous solid chunks and powders, wood and biomass, plastics, RDF, MSW, etc.

C. *Rolling Bed Dryer*

In a rolling bed dryer, the material is circulated and mixed by rotating paddles. The drying air is supplied through a perforated plate. This dryer has combined features of a fluidized bed and rotary dryer. These are often used for drying wood chips and organic residues. These are used for drying biomass and recycling, and in the manufacture of wood particle board, pellet, and biofuels.

D. *Spray Dryers*

Spray drying is a method of producing a dry powder from a liquid or slurry by rapidly drying with a hot gas. The hot gas can be waste heat gas or heated air. If the material to be dried has flammable content, then heated nitrogen also can be utilized. Spray dryers use an atomizer or spray nozzle to spray the liquid or slurry. The most common spray dryers are single effect ones. To overcome the dust and flow issues of the powder, multiple effect spray dryers are utilized.

Industrial liquid waste streams, slurry waste streams, are generally dried in the spray dryers.

E. *Conduction Dryers*

Conduction drying are contact dryers, and they are suitable for wet particles as their thermal efficiency is higher. The evaporated water vapour or organic solvent is extracted using vacuum. Vacuum operation is very useful in heat-sensitive granular materials. Some of the important types of conduction dryers are the following:

(i)   Hollow blade dryer;
(ii)   Vacuum rake dryer;
(iii)   Belt drier;
(iv)   Drum scraper dryer;
(v)   Double-cone rotary vacuum dryer.
        The conduction dryers are used for pasty or heat-sensitive industrial hazardous and non-hazardous materials.

F. *Convection Dryer*

The convection dryer consists of a moving perforated belt which is travelling inside an enclosed casing. Material to be dried is dropped on this belt, and the hot gas is passed through the belt and feed. The belt moves and a seal between the static (stationary) and dynamic (moving) components contains the bed, preventing short-circuiting of the carrier gas. The heat source, belt, drives, feeding mechanism, and primary gas movers are installed in a frame, and the entire system is insulated. Entry doors along the length and ends of the dryer provide access to the moving components. Heat-sensitive industrial hazardous and non-hazardous materials can be dried on the same.

11.3.3 Impregnation

In the impregnation process, the sludge or viscous material which is difficult to transport using a pump or conveying equipment due to its sticky nature is made to mix with dry organic or inorganic materials so that it loses its sticky nature and becomes flowable.

Figure 11.7 depicts the impregnation process of the hydrocarbon sludge. In this operation, sludge from drums is dropped into a pit containing dry powder. The contents in the pit are thoroughly mixed using a mechanical device.

Depending upon the volume of material and the economics of the operation, impregnation operation can also be implemented in a completely automated manner through remote control.

The impregnation of sludges or pastes with dry materials can be carried out by mixing these materials on the floor or in a pit. The mixing can be performed manually using different equipment such as a spade, payloader, and arm handler. After achieving the desired level of mixing of the pasty material and the dry impregnation material, this mixture is sent through other equipment such as extruder, blender, and sieving to achieve AFR having uniform physico-chemical characteristics.

Sludges from the oil industry, tank bottom sludges, ETP sludges from chemical industries, out of specification or out of date paints, oils, chocolates, juices and other similar materials, distillation residue, failed batches of sticky materials, etc. are impregnated using different impregnation materials such as sawdust, rice and other husks, dry inorganic materials such as limestone powder, coal powder, raw meal of cement process, ESP or baghouse dust, and other inorganic powdery materials.

Fig. 11.7 Drums with hydrocarbon sludge are emptied in a mixing pit

11.3.4 Bailing

Bailing is an operation to densify the loose material to reduce its storage footprint and transportation cost. A bailer converts loose material into blocks of different geometry. Various kinds of manual, semi-automatic and automatic bailing machines called bailers are available in the marketplace. The automatic bailing machines may be driven electrically or hydraulically. Bailing machines are used for creating compact bundles of materials such as paper, plastics, cardboards, tyres, metals, agro-waste, processed waste, RDF, and dry MSW. Figure 11.8 shows a typical bailing machine.

11.3.5 Segregation

Segregation is a very important operation in segregating constituents present in MSW, RDF, mix plastic waste, mix paper waste, e-waste, C&D waste, and other kinds of wastes. A new approach which is gaining popularity is to engage robots in the segregation process where robots can be programmed to segregate waste materials based on the shape, weight, colour, appearance, text, logo, photo, etc., defining the products.

Segregation of wastes into different constituents can be implemented in a manual or automatic mode. These segregation processes are depicted in Figs. 11.9 and 11.10.

Fig. 11.8 Typical bailing machine

Fig. 11.9 Manual Sorting Process

A. *Manual Segregation Process:*

In manual segregation, the waste is segregated into different fractions by involving manpower. To facilitate efficiency in this process, generally, a sorting conveyor is implemented on which the material moves and during this movement, the waste of different kinds is picked up by the engaged personnel.

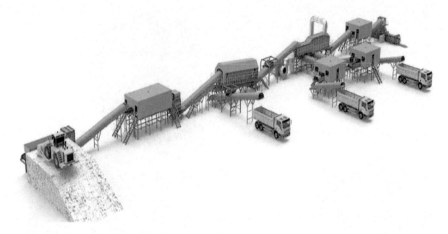

Fig. 11.10 Automatic sorting machine

B.   *Automatic segregation Process:*

Automatic waste segregation machine makes use of a variety of sorting means to separate organic matter, plastics, metal, bricks and stones, and other substances out from the mixed waste to the maximum, to improve the reusing and recycling of waste. At the same time, the separated waste materials can be further re-processed into useful resources. So, the main purpose of the automatic waste sorter is reduction processing and turning waste into treasure.

Various technologies get employed in the automatic segregation process by utilizing different physical properties such as density, colour, shape, size, nature of material, conductivity, and others.

11.3.6   Blending

Blending refers to the mixing of ingredients to achieve the desired property to the mix. As elaborated in the introduction of this chapter, waste does not have any quality consideration. It properly varies from source to source, time to time, lot to lot, etc. When wastes must be utilized as resources, they need to have uniform quality. To achieve uniform physical–chemical characteristics in the waste, it needs to be processed by various pre-processing operations defined above. In addition to implementing the above operations, for waste to have uniform characteristics, necessary ingredient needs to be incorporated in the same. This objective is achieved through blending.

Whether wastes are liquids, solids, or sludges, blending operation is very much relevant for converting them into AFRs. Blending operation is carried out to modify the calorific value, ash content, moisture content, chlorine content, and other parameters of the wastes to the desired uniform level.

Blending operation is carried out in different modes. For example, in liquid streams, blending is carried out in agitated tanks, for solids it is carried out in the shredding operation, and in sludges it is carried out during the impregnation process. To ensure uniformity in the processed waste stream, at least one lot is prepared and kept ready for use and the other lots remain in preparation mode.

Various waste streams that are processed using blending are aqueous liquid streams, organic liquid streams, RDF, plastic wastes, industrial hazardous wastes, industrial non-hazardous wastes, agro-wastes, etc., and also a mix of all of these different kinds of wastes.

11.4 Waste Acceptance Criteria

The following criteria need to be respected while accepting the waste streams for pre-processing or co-processing:

a. They should not influence the cement quality in terms of setting times or the early strength. (AFRs having phosphate levels higher than the prescribed limits.)
b. Their use in cement should not lead to leaching of the heavy metals from the concrete. (AFRs having more heavy metals than clinker capacity to handle them. Many countries prescribe the limits of different heavy metals in the AFR.)
c. They should not impact the emissions or cause damage to the environment. (Materials containing Hg, hexavalent chromium, etc.)
d. They are safe to handle and process in the given facility. (AFRs having lower flash point than the flash point for which the facility is designed.)
e. They do not impact the cement production process. (AFRs having a higher level of chlorine, alkalis, or sulphur than the process can handle.)

11.5 Salient Features of the AFR Pre-processing Facility

The typical pre-processing platform suitable for the solids consists of the following important features.

11.5.1 Entry Gate

The entry gate is provided to ensure that the material entering the pre-processing facility is meeting the required statutory, commercial, environmental, safety, and administrative protocols. These need to be checked before accepting the waste/AFR inside the plant.

Table 11.1 Parameters for testing AFRs

S. No.	Parameter	Remarks
1	Physical State Solid Liquid Sludge	Size Viscosity Viscosity
2	Proximate Analysis	Moisture, Ash, Volatiles, Fixed Carbon
3	Ultimate Analysis	Carbon, Hydrogen, Nitrogen, Sulphur
	Ash Composition	CaO, SiO2, Al2O3, Fe2O3, K2O, Na2O, MgO, P2O5, etc.
	Net Calorific Value (NCV)	Kcal/kg, MJ/Kg
	Density	Kg/m3
	Halogens	Chlorine, Bromine, Fluorine
	Flash Point	Deg. C
	Heavy Metals	Sb, As, PB, CO, Cr, Cu, Mn, Ni, V, Cd, Hg, Tl, Zn, Be

11.5.2   Weigh Bridge

Weigh bridge is desired to ascertain the weight of the incoming and outgoing material.

11.5.3   Laboratory

The laboratory is implemented to evaluate the quality of the incoming wastes, intermediately processed materials, and fully processed AFRs. The instruments and equipment required for the assessment of the materials depend upon the nature of the incoming material, intermediately processed material, and fully processed AFRs.

The various parameters that need to be analysed are provided in Table 11.1.

Also, when AFRs/waste streams are received at the plant, their acceptance based on fingerprint analysis or detailed analysis also is required to be carried out.

It is also important to find out the solvents in which they are miscible such as oil, solvents, and water. Based on the properties of the AFRs and the prescribed limits of the relevant constituents, it is desired that their minimum, maximum, and average feasibility of co-processing are estimated and kept ready.

11.5.4   Storage Shed

Storage shed for appropriate storage of incoming material, intermediately processed material, and fully processed AFRs.

11.5.5 Pre-processing Plant and Machinery

Pre-processing plant and machinery consists of different plants and machinery designed and implemented to process the defined waste streams into AFR. It also contains the equipment required to handle, move, and transport the waste materials, intermediate materials as well as the processed AFRs. The design of the entire plant and machinery for pre-processing would depend upon the physico-chemical nature of the waste material (LafargeHolcim & GIZ, 2020).

A.   Solid Wastes

The plant and machinery required for pre-processing of solid wastes consists of a properly sized shredder. For removing ferrous metals a magnetic separator and for non-ferrous metals, an eddy current separator are used. The inert materials are removed by implementing the pneumatic separation facility, and oversized materials are removed by screening through an appropriate screening arrangement. Oversized material is sent for recycling back to the shredder.

The solid waste streams can also be utilized for impregnation purposes depending upon their characteristics. Sawdust, rice husk, other biomasses, spend carbon, paper waste, cloth waste, etc. are examples of the same.

B.   Liquid Wastes

The plant and machinery for pre-processing of liquids consists of an agitated tank with a screening arrangement of the coarse and fine particles. Before feeding the new liquid stream in the tank, a compatibility test is carried to check whether a new volume is compatible to mix in the existing liquid stream present in the agitated tank.

C.   Wastes Sludges

The plant and machinery for pre-processing of sludges consists of a floor or pit-based impregnation arrangement. The mixing of the sludge with dry material is made using an appropriate tool and then is passed through suitable mixing equipment.

Sludge wastes can also be mixed with liquid wastes to reduce their viscosity and fed as liquid AFRs by pumping them using suitable pumps.

Sludge wastes can also be pumped directly into the kiln inlet or riser duct using a concrete-type pump.

Pre-processing facilities facilitate the conversion of different quality wastes into uniform quality material that can be fed smoothly in the kiln system. Therefore, these facilities must be able to accept wastes in different packaging types and handle different kinds of wastes having varying properties. The facilities should be able to convert different wastes into AFRs that can be fed through the available feed points. It is important to ensure that the facility is designed in such a way that it can expand its capacity smoothly in the future.

11.5.6 Environmental and Safety Provisions

As wastes are handled in the pre-processing facility, several environmental and safety precautions need to be considered depending upon the nature of the material. They are deliberated below.

A. Environmental Provisions

Volatile organic carbon emissions, floor seepage of hazardous or non-hazardous waste streams, and odour are the three important environmental considerations that need to be addressed while designing the facility.

B. VOC emissions

To deal with the VOC emissions, these may be exhausted from the facility, and then they are either thermally destroyed or catalytically destroyed or adsorbed on a suitable adsorbing media such as activated carbon. If the pre-processing facility is installed within the cement plant, then these exhaust gases can be diverted into the kiln.

C. Floor spillages of the waste streams

They need to be managed by having an impervious floor in the facility so as to avoid its seepage in the soil and groundwater. The impervious floor can be achieved by installing a geopolymer membrane below the concrete flooring. Further, the facility needs to have a proper drainage system to divert the seepages into a containment tank for subsequent treatment through a suitable technology.

D. Odour Control

Most of the waste materials have an undesired odour which causes substantial concern to the employees and the surrounding community. The VOC emission control may reduce the intensity of the odour but to improve the situation further, there are two more technology options. The first one is the use of fragrance spray in the premises and the second is the use of adsorbing sheets installed in the direction of the wind so that the VOC emissions are adsorbed onto these sheets.

E. Safety Provisions

Exposure to toxic materials, runaway reactions due to incompatible materials, fire and explosion due to various reasons such as flammability of the materials, low flash point-bearing materials, inappropriate design of the machinery and electrical facilities from ATEX considerations, etc. To deal with this situation, the following considerations need to be considered.

F. Equipment design

The equipment needs to be certified to the prevailing ATEX conditions in the facility.

G. Exposure to toxic materials

Proper exposure protection needs to be ensured using appropriately selected personal protective equipment (PPEs).

H. Runaway reactions due to incompatible materials

The compatibility study of the different materials needs to be properly studied/understood and the incompatible materials need to be stored away from each other to avoid their contact. The storage of the material needs to be properly designed in the pre-processing facility accordingly.

I. Electrical system design

The design of the electrical infrastructure in the pre-processing facility must meet the compliance requirements depending upon the hazard characteristics of the materials handled in the facility.

J. Fire prevention and control

The facility needs to be equipped with appropriate fire prevention and control arrangements to deal with the fire in case it unfortunately occurs. The design of the same must comply with the local statutory requirements. If relevant, to deal with the contaminated used firewater, the same needs to be collected in a pond for subsequent evaluation and treatment. One of the most appropriate options for managing the floor spillages and contaminated firewater is to dispose of it in cement kiln through co-processing.

11.6 Other Pre-processing Options

Instead of pre-processing the waste streams in a pre-processing facility using various unit operations mentioned above, it may be feasible to implement options of treating the waste streams in external combustors, pyrolysis units, and gasifiers. In these operations, the waste gets combusted into this treatment equipment and the hot gases then enter the kiln. These treatment facilities are reasonably fine for lump fuels. They help to control the chlorine and ash content—present in waste streams—getting into the kiln and causing problems.

11.7   Production of Alternative Fuels by Pre-processing and Supply to Cement Plants

For successful pre-processing, it is important to have the analysis of the physico-chemical characteristics of all the waste streams available in the stores. The cement kiln requires AFR to have certain desired characteristics. AFR having such desired characteristics is produced by utilizing a mix of these waste streams in defined proportions and utilizing different waste treatment technologies that are illustrated above.

While arriving at the desired proportion mix, the solid, sludge, and liquid AFRs need to be properly configured and they need to be pre-processed accordingly and blended to arrive at the final AFR mix that meets the acceptable AFR quality as desired by the cement kiln. For this, therefore, the solid, sludge, and liquid AFRs need to be suitably pre-processed by utilizing different technologies as defined above.

11.7.1   Many of the waste management companies pre-process hazardous and non-hazardous wastes into alternative fuels and supply them to cement indus-tries. Pre-processing of wastes into alternative fuels is practiced successfully in many different countries and large-scale production of alternative fuels is carried out. The alternative fuels are produced in both liquid and solid states and co-processed by cement plants. In India also, this trend has started and a few of the waste management companies are converting wastes into AFRs and providing the same to cement kilns for co-processing.

11.7.2   Figure 11.11 provides a typical demonstration of the AFRs being supplied by Green Gene Enviro Protection & Infrastructure Ltd. (GEPIL), India, which is a large supplier of pre-processed AFRs (www.gepil.in).

Figure 11.12 provides details of another supplier of pre-processed alterna-tive fuels—BEIL Infrastructure Ltd. from India who are also the suppliers of pre-processed AFRs in India.

11.8   Pre-processing of Different Kinds of Wastes into AFRs

The following flow sheets and deliberations provide inputs on the design and imple-mentation of the pre-processing systems for solid wastes, liquid wastes, and waste sludges.

Fig. 11.11 Supply of Pre-processed Alternative Fuels by GGEPIL, India

11.8.1 Pre-processing of Solid Wastes

Pre-processing of solid wastes is carried out by implementing unit operations such as blending, segregation, shredding, and drying. A typical flow sheet of this system is provided in Fig. 11.10.

Pre-processing of solid wastes consists of the following steps:

- The first step is to test the inbound material received from the market and accept it for storing in the storage shed based on the results of the evaluation.
- Different materials having different physico-chemical properties are stored separately.
- The composition of the batch mix of different solid waste streams available in the stores to achieve the desired uniform characteristics of the AFR, based on the characteristics of the individual streams, is determined.
- Then the shredding operation is started, and the waste materials are fed in the hopper of the shredder as per the batch mix composition using a payloader.
- The output of the shredder then is conveyed to a wind sifter. While this conveying is in progress, the metal detector and the eddy current separator remove the ferrous and non-ferrous metal fractions from the shredded material.

Pre – Processing Facility at BEIL Infrastructure Limited
(www.beil.co.in)

BEIL Infrastructure Limited (BEIL) is located at Ankleshwar, Gujarat, India. BEIL is a pioneer in India for TSDF since 1997 and is ISO 14001 & ISO 45001 certified. From 1997 till date, BEIL has ventured in Common Incineration, Common MEE, Pre-processing, Plastic waste management etc. BEIL started its co-processing activity in 2016 and its first collaboration was with M/s Ambuja Cement, Junagadh, Gujarat. At BEIL all the solid and liquid waste received is evaluated for co-processibility and is diverted to the pre-processing facility. In the pre-processing facility, Solid and Liquid wastes are handled separately ensuring utmost safety in operation. Facilities like mixing tanks, blenders, scrubbers are available to ensure least diffusion in the atmosphere.

Liquid waste received at pre-processing facility in drums are checked for compatibility within each other and also with material already available in mixing tank. The compatible liquid in drums are then emptied in the mixing tank. After mixing, and testing for compliance to quality standard is filled in tanker for transport to cement industry for co-processing. Solid waste received at pre-processing are checked for compatibility to prepare the blend with other evaluated materials. In case required, saw dust/fly ash is added to it to make it flowable. This material is then added to the existing waste mix and blended properly. After blending and testing its compliance to quality, it is filled in bags and sent for co-processing.

Every year BEIL sends around 10000 MT of solid waste mix and 12000 MT of liquid waste mix for co-processing at cement industries located in various states like Gujarat, Chhattisgarh, Rajasthan, Karnataka& Madhya Pradesh. We are also exploring more such facilities.

Fig. 11.12   Supply of pre-processed Alternative Fuels by BEIL, India (www.beil.co.in)

- Subsequently, the wind separator segregates heavy fraction such as stones and grit from the light combustible material.
- The light combustible material is then screened on a police screen to segregate the desired size fraction and the oversize fraction.
- Desired size fraction of the shredded material is then stored in the form of a heap. From this heap, prepared AFR material is sampled and tested to confirm that the desired characteristics are met.
- In case the same are not met, then the necessary corrective materials are added to this AFR material in the heap and blended properly using the payloader. Subsequently, the ready AFR material is shifted to the outbound storage area for sending for co-processing.
- The oversize fraction from the police screen is sent back to the shredder using the oversize return conveyor.

Pre-processing of Liquid Wastes

Pre-processing liquid wastes are carried out by implementing unit operations of blending. A typical flow sheet of the same is provided in Fig. 11.13.

Figure 11.14 depicts the pre-processing aspects related to liquid wastes.

The following are the various steps involved in the pre-processing of the liquid wastes.

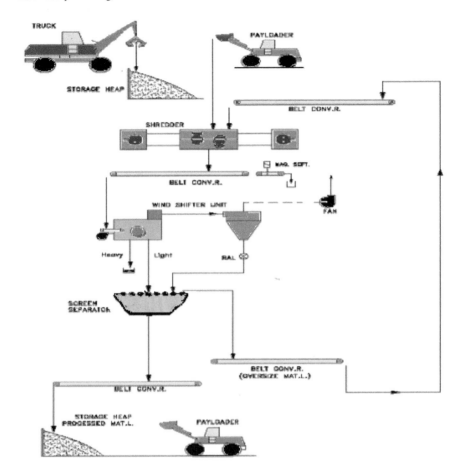

FLOW SHEET FOR PRE-PROCESSING OF SOLID WASTE

Fig. 11.13 Typical flow sheet for the pre-processing of solid wastes

- Liquid pre-processing facility consists of installation of different storage tanks with appropriate unloading pumps as shown in Fig. 11.15. For storing the waste, which is in the form of a slurry, an agitated tank is provided.
- The required safety systems are provided on the respective tanks depending upon the physico-chemical characteristics of the waste streams.
- Whenever a new tanker of liquid waste is received, its nature and characteristics are evaluated by carrying out the relevant laboratory tests.
- Based on the characteristics, the tank to which this tanker has to be unloaded is decided.
- Once the tank to move the new material is finalized, a sample of the material from that tank is taken and a compatibility test of the new material received in the plant

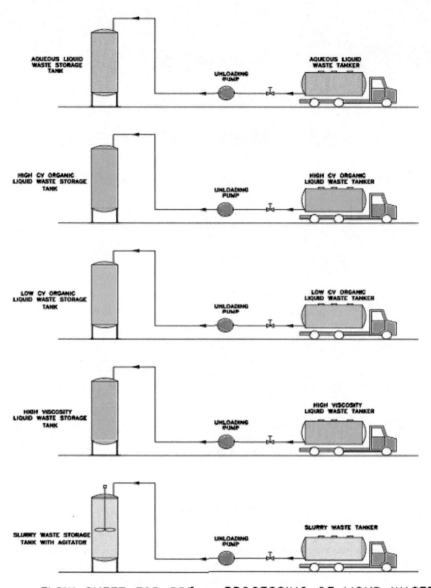

FLOW SHEET FOR PRE — PROCESSING OF LIQUD WASTE

Fig. 11.14 Typical flow sheet for the pre-processing of liquid wastes

Fig. 11.15 Container arrangement for sludges

along with the old material present in the tank is carried out using the defined procedure.

- After it is confirmed that the material to be added to the tank is compatible with the one stored in the tank, then the new received material is unloaded in that tank using the unloading pump.
- In case the waste is in the form of a slurry, then the same is stored in an agitated tank.
- When the liquid streams are required to be co-processed, depending upon their compatibility, two or more waste streams from different tanks can be mixed in line and sent to the calciner or main burner for co-processing.

11.8.2 Pre-processing of Sludges

The following are the various steps involved in the pre-processing of the sludges.

The sludge waste is transported in drums or different kinds of containers as shown in Fig 11.15.

Pre-processing sludges is carried out by implementing unit operations of impregnation, blending, etc. A typical flow sheet is provided in Fig 11.16.

- Sludge pre-processing facility consists of installation for mixing of the sludge with a dry powdery material such as sawdust, rice husk, other biomasses, limestone powder, raw meal, and ESP or baghouse dust.
- After receipt of this sludge at the plant, it is tested for the required properties and is then shifted to the stores.
- The sludge present in drums or the containers is then dropped in a pit or a sludge mixer.
- Required quantity of compatible dry powder is then dropped in the pit or the mixer. These dry powders can be sawdust, rice husk, and other biomass powders that can be procured from the market. These can also be limestone powder, raw meal powder or ESP/baghouse dust, etc., which is available in the cement manufacturing plant.
- The sludge and dry powder are thoroughly mixed to make the sludge non-sticky and hence flowable. This process is called impregnation.
- The impregnated sludge prepared as above is then tested for its characteristics and sent to a storage shed.

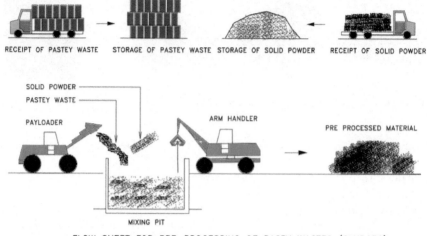

Fig. 11.16 Typical flow sheet for Pre-processing of Pasty Wastes

The pre-processing facility is designed as per the specification of the waste materials to be handled, stored, and pre-processed. These are designed and supplied by different pre-processing system suppliers.

Figure 11.17 provides a view of the pre-processing facility supplied by Sanghavi Engineering Pvt. Limited, India.

11.9   Management of Drums and Other Packaging Types

The various packaging in which AFRs will get received in the pre-processing facility/co-processing facility would be dependent upon the type of AFR. These are listed in Table 11.2.

The wastes/AFRs being brought in these packaging could be hazardous or non-hazardous in nature. In the case of plastic or cloth-type packaging, irrespective of the contamination with hazardous or non-hazardous material, the same can be shredded to an acceptable size and sent to the cement kiln for co-processing.

The metal drums that contained non-hazardous material can be vacuum emptied and reused for storing similar materials or can be sent to the scrap dealer.

In case the contamination is hazardous in nature, then the same needs to be sent to an authorized drum treatment facility as such or after pressing them hydraulically.

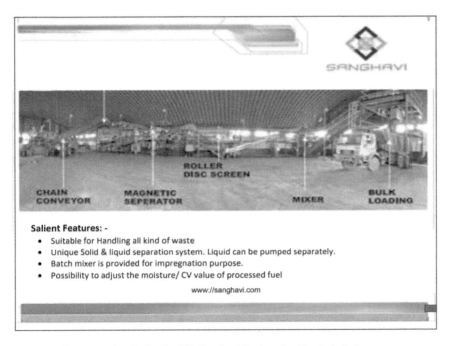

Fig. 11.17 Pre-processing facility by M/s Sanghavi Engineering Pvt. Ltd., India

Table 11.2 Types of packaging of waste/AFR materials

Solid	Plastic or cloth bags of different sizes, loose, etc.
Liquid	Plastic cans of different sizes, plastic drums of 200 lit capacity, metal drums of 200 lit capacity, 1000 lit intermediate bulk container, truck-based tankers
Sludge	Plastic drums 200 lit, metal drums 200 lit, sludge containers

11.10 Conclusion

Pre-processing of wastes into AFRs is an important consideration in achieving successful co-processing of AFRs. Wastes, by their very definition, do not have any quality consideration. They have lot to lot variation in their physical and chemical characteristics. If such wastes are fed into the kiln in the same manner as they are available, they cause disturbance in the kiln process impacting product quality as well as performance parameters of the kiln. Hence, wastes need to be processed to achieve uniform physical and chemical characteristics. This processing involves several unit operations depending upon the nature of wastes. These unit operations include blending, shredding, drying, impregnation, size separation, segregation, bailing, etc. This chapter has evaluated various aspects associated with pre-processing of wastes into AFRs.

References

LafargeHolcim & GIZ. (2020). Guidelines on Pre and Co-processing of waste in cement production.
www.gepil.in
www.beil.co.in
www.loesche.com
www.wasteconditioning.com

Chapter 12
Operational Considerations in Co-processing

12.1 Introduction

Cement kiln co-processing operation is required to respect and deal with various concerns that are encountered while undertaking co-processing. These concerns are related to various aspects such as sustainability, legal, environmental, technical, quality, health and safety, transparency in communication, etc. These concerns need to be addressed through appropriate operational assessments, procedures, and practices. The same are discussed in detail below.

12.2 Operational Guidelines for Co-processing

Following are the various operational considerations that need to be taken into account while undertaking co-processing in cement kilns.

12.2.1 Sustainability

It is important that co-processing respects the waste management hierarchy and it should not hamper the waste reduction efforts. Hence, waste should not be co-processed if ecologically and economically better ways of treatment are available.

12.2.2 Legal Aspects

Co-processing shall be carried out in line with the legally binding manner so as to assure a high level of environmental protection. It is important, therefore, to identify all relevant laws, regulations, standards, and company policies relating to safety, health, environment, and quality control. Appropriate equipment and/or management procedures need to be put in place to ensure complete compliance with them. All employees and contractors need to be made aware of the relevant laws, regulations, and standards, and also made to understand their responsibilities under them.

12.2.3   Acceptance Process of Wastes and AFRs

Operators should develop appropriate acceptance procedures for wastes and AFRs. This must include the following:

- Quality assessment of the waste materials and AFRs available from agencies with due consideration to the chlorine, sulphur and alkali content, phosphate content, water content, heat value, ash content, heavy metals content, volatile content, etc., along with an assessment of their impact on clinker and cement quality.
- Waste safety assessment with data sheets documenting the chemical and physical properties, health, safety and environmental considerations to be followed during sourcing, transportation, handling, storage, pre-processing, and co-processing. Understanding of the PPE.

12.2.4   Manpower

Adequate number of manpower and skills are required for safe and successful co-processing operation. An appropriate process needs to be in place for their training and retraining.

12.2.5   Operation and Management

Appropriate manuals and SOPs for the operations and management of the pre and co-processing facility. Different kinds of wastes and AFRs needs to be properly documented and put in practice. The linkages related to co-processing operations and kiln start up, shut down, upsets, etc., are properly documented and put in place.

12.2.6   Emergency Management

Use of wastes and AFRs may raise many emergency situations. To deal with these situations, it is important to keep ready an Emergency Management Plan and a well-trained response team.

12.2.7 External Communications

Co-processing of waste and AFRs require that a strong Stake Holder Engagement Plan is put in place to ensure that the community is aligned and engaged to co-processing and it happens in a smooth manner.

12.2.8 Design Considerations

The design of equipment, systems, and facility has to be compatible with the material characteristics so that the desired level of success is achieved in co-processing the same without any untoward impact on the environment. It is also important to keep their documentation revised and up-to-date.

12.2.9 Material Receipt and Storage

Following are the important aspects that need to be ensured while receiving and storing the waste and AFR materials:

i.   All the materials are received based on the permits granted by the relevant authorities.
ii.   All the received materials are properly evaluated for their physico-chemical characteristics.
iii. The storage of material should ensure that incompatible materials are kept separately and securely.
iv.   The storage facility is adequately designed to deal with the spillage, fire, dust, odour, and other similar risks that are encountered due to the storage of waste and AFR materials.
v.   The responsible personnel is properly trained to deal with the above defined risk factors and also the emergency situations.

12.2.10 Material Handling and Feeding Systems

The material handling & feeding systems have to be compatible with the physico-chemical characteristics of the waste and AFR materials. While designing these systems, it is important to consider all the risks such as fugitive emissions, odour, health and safety issues, etc. The personnel handling these systems need to be trained properly in all the relevant aspects of the materials and also the equipment. They should also be trained in dealing with the emergency processes.

12.2.11   Material of Non-Compliant Deliveries

There should be a written procedure to deal with the material that is non-compliant with the agreed specifications and is also shared with the suppliers. Where ever relevant, the communication with respect to rejection of non-compliant material needs to be sent across to the relevant authority. In case required, there has to be a procedure in place to send it for alternative treatment option in case it is not feasible to return the same back to the waste generator.

12.2.12   Quality Control

A proper procedure must be defined for ensuring quality control of all the materials received, pre-processed or produced at the facility. This procedure includes detailed instructions pertaining to the following:

- Sampling & analysis,
- frequency of sampling,
- laboratory protocols and standards,
- Analysis, calibration procedures, and maintenance,
- recording and reporting protocols.

An adequately designed laboratory facility must be in place with the required infrastructure and testing equipment. It should be operated by adequately trained personnel.

12.2.13   Process Control for AFRs

AFRs need to have a constant quality and feed rate so as to ensure smooth kiln operation as well as good product quality. It is important, therefore, to monitor inputs, processes, products, and emissions, so that proper operation of the kiln is maintained. The kiln emissions need to be monitored online. Many countries including India have mandated the installation of Continuous Emission Monitoring System (CEMS) on the kiln stacks and hooking them to the server of the pollution control boards. It is also important to ensure that proper controls are exercised to achieve good control over the kiln process. AFRs are not to be co-processed during start up, shut down, when the temperature in the kiln is not at the operating levels, and when the emission monitoring system is not operating. The kiln operators need to be trained to operate kilns while co-processing of AFRs in a gradual manner.

12.2.14   *Management of the Co-processing Activity*

The facility must have written procedures and instructions for unloading, handling, storage of the materials at the site. All the personnel must be adequately trained in their respective trades and skills. The facility is operated in such a manner that there are no concerns pertaining to HSE and Quality of the products. The routes for the entry and exit vehicles are properly defined, appropriate signages are placed around the facility and also near the stored materials. The selection of feed point for co-processing of waste material or AFRs needs to be done based on the nature of the material and also as per the statutory mandates if any. Materials having chlorine content > 1% are desired to be co-processed through the main burner to avoid the formation of toxic compounds such as Dioxins/Furans. Safety, process, fire, emergency, and all statutory audits are conducted regularly at the facility to ensure preparedness. Security monitoring at the facility is also an important consideration so as to ensure avoidance of unnecessary trespassing of outside people. Appropriate communication with the community on the ongoing co-processing operations at the plant is also an important consideration that needs to be carried out through well-designed communication plan. Engagement with the community in the waste and AFR activity is an important way to align it towards co-processing. Procurement of non-cattle feed biomass, plastic & other non-recyclable combustible waste materials from the community in a business mode is a very powerful tool to build a strong bond with the stakeholders.

12.3   Technical Considerations for Successful Co-processing

The chemistry of the wastes and AFR materials is different in some respects than the conventional natural materials. For example, wastes will have higher quantities of Sulphur, chlorine, phosphorous, heavy metals, minor elements, alkalis, other oxides, etc. Also, their physical characteristics are totally different than the conventional materials. Hence, we tend to observe new and different issues while co-processing the AFRs. It is important, therefore, to take cognizance of the physico-chemical characteristics of waste and AFR materials. Accordingly, it is required to design the co-processing operation and monitor and control relevant parameters to the desired levels to achieve successful co-processing. These aspects are illustrated below.

12.3.1   *Raw Mix Design*

Raw mix design is a critical step in the manufacture of Cement. To achieve the desired quality of cement, it is necessary to have appropriate proportioning of the raw materials used in the cement process. For the raw material proportioning to be appropriate, their accurate chemical analysis is very critical. While carrying out the raw mix design, the ash analysis of the fossil fuel is also considered appropriately.

When AFR is utilized to substitute the fossil fuel or fossil raw material, its ash analysis is also very important—especially when the proportion of AFR utilization (TSR%) is high and increasing. Generally, AFR is prepared by pre-processing many different waste streams. Hence, the design of AFR mix needs to be carried out by using a detailed analysis of the different waste streams being part of the AFR mix. Subsequently, the detailed analysis of the AFR is utilized in designing the raw mix in cement manufacture.

For achieving the desired quality cement and also a stable process operation, a certain raw mix formulation is designed and implemented. It is important to note that the same raw mix design needs to be achieved even after incorporating AFR in the fuel mix. Hence, this step needs to be added to the routine raw mix design process and followed.

12.3.2   Alkalis, SO₃, and Chloride Balance

Alkalis, chlorine, and sulphur reduce kiln efficiency in the cement industry. Their compounds like alkali chlorides, alkali sulphates, calcium sulphates, etc., are thermally unstable at high temperatures and they get vaporized or decomposed. These volatilized compounds condense back in the preheater system and return with the solids to high-temperature zones. The continuous volatilization–condensation reactions cause cycling and also create coatings of these compounds. The stability of the kiln process gets reduced in the process and causes kiln stoppages. To avoid such consequences, it is important to monitor the alkalis, chlorine, and sulphur levels in the hot meal sample on a frequent basis. This frequency of monitoring depends upon the extent of impact imparted by these constituents on the cement manufacturing process.

There are two different criteria with which the impact of alkalis, chlorine, and sulphur can be evaluated. The first one is the concentration of chlorine and SO_3 in the hot meal and the other is measuring the alkali sulphur balance in the hot meal.

i.   *A/S Ratio*:

The following equation is used to evaluate the alkali/sulphur (A/S) ratio (CII, 2010).

$$A/S = \frac{\{(K_2O/94) + (Na_2O/62) - (Cl/71)\}}{SO3/80}.$$

It is desired to control the same between 0.8 and 1.2. A/S values above or below this range require a high level of operating controls.

Alkali chlorides are far more volatile than alkali sulphates and recirculate within the kiln. Hence, in the hot meal sample, K_2O or Na_2O tied up with the chlorides are not considered in the A/S ratio calculation. The hot meal A/S ratio predicts the likelihood of alkali or sulphur related build-ups in the kiln inlet. A high A/S ratio indicates the portion of the alkalis which do not combine with SO_3 and recirculate in the kiln. This provides increased potential for the formation of rings and preheater build-ups. A sudden decrease in the A/S ratio indicates a lack of oxygen in the kiln end causing sulphur build-ups.

If alkalis are very high and are not balanced by sulphur, they will continue to recirculate within the kiln/preheater system, this increases the possibility of the formation of kiln rings and preheater build-ups. The same impact is seen when sulphur is high and is not balanced by alkalis. Excess sulphur in the hot meal can also form sulphospurrite ($2(CaO) SiO_2CaSO_4$). It forms exceedingly hard and dense build-ups. Clinker quality would also suffer because sulphur which is not combined with alkalis forms a solid solution with the silicate minerals. These results increase the C2S content and decrease in C3S content in the clinker, thereby reducing the cement strength.

ii.   *Chlorine & SO3 Concentration:*

Figure 12.1 graphically illustrates the methodology of assessing the impact on the kiln process by evaluating the concentration of chlorine and SO_3 in the hot meal (Montes de Oca & Forinton, 2010).

It is desired to control the concentration of chlorine and SO_3 in the hot meal in the ranges provided in Fig. 12.1 to ensure that the coating issues are avoided. Chloride will combine with all of the alkalis present in the kiln forming alkali chlorides if the same is very high in concentration. It will, therefore, keep recirculating in the kiln with increased possibilities of build-ups in the preheater. The remaining chloride will form $CaCl_2$ by combining with CaO. $CaCl_2$ has a very low melting point (770–780°C). This makes the hot meal very "sticky" at these temperature levels. This sticky material, therefore, increases the possibility of build-ups at the higher end of the preheater. Chlorides also form eutectic mixtures when they react with sulphates

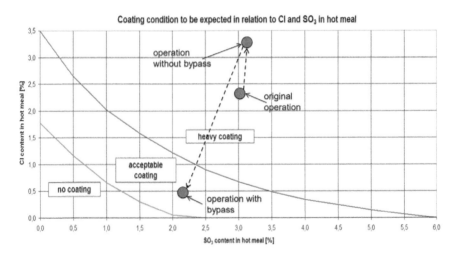

Fig. 12.1 Guideline for identifying the critical concentration of Chlorine and SO_3

of potassium, sodium, calcium, and magnesium. These eutectic mixtures have lower melting points than that of the pure compounds. This further increases the likely hood of rings and build-up formations.

It is important to understand that it is possible to deal with the values of A/S ratio that are substantially higher or lower than those mentioned above and also the Chlorine and SO_3 concentration that are way higher than those depicted in Fig. 12.1 as safe to operate. For operating kiln in these conditions, various other techniques are employed which include incorporation of meal curtain, installation of mechanical hammers to dislodge the coating, employing advanced features of kiln process control, etc.

12.3.3 Chlorine Limits in Clinker

The control of chlorine content in clinker is an important parameter to achieve smooth operation of the kiln for clinker manufacture. In the earlier days, these limits were getting defined based on the presence of chlorine content in the raw material. The conventional norm for chlorine limit in clinker is 300 g/T of Clinker. Hence, it is important to monitor the Chlorine coming from RM, conventional fuel, and also AFR, so that the chlorine in the kiln system is controlled at the set limit. Hence, monitoring of the same in these materials on a regular basis is an important requirement. The monitoring of chlorine in the hot meal also is critical as mentioned in the earlier section above.

It should be noted that the kiln can be run at higher chlorine contents than the limit of 300 gm/T cl. However, as mentioned in the earlier section, additional and specific measures are required to be implemented to mitigate the impact of higher chlorine content on the kiln process.

Figure 12.2 provides a case study of dealing with higher levels of Chlorine and SO_3 (Lowes & de Souza, 2017). When the chlorine quantum rises beyond the acceptable limits, then a chlorine by-pass needs to be installed to remove the chlorine from the system. The chlorine by-pass consists of removing a part of the gases.

12.3.4 Fuel Mix Design

Several waste materials are used in the kiln to manufacture the clinker. The chemical characteristics of all these waste materials are substantially different from one another as well as may contain constituents at levels that may not be acceptable in the process. Further, AFR, which is prepared by pre-processing all different kinds of waste materials, must meet certain input standards so as to achieve desired quality clinker. Hence, the AFR chemistry must be compatible with the chemistry of the raw materials and the conventional fossil fuel that is used in the plant. The various parameters associated with the acceptance criteria of the AFR include calorific value, moisture

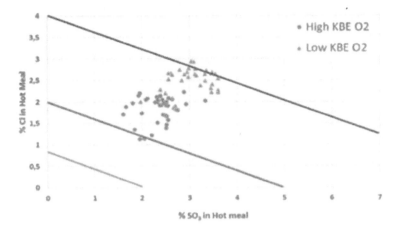

Fig. 12.2 Operations with additional quantum of Chlorine and SO$_3$

content, ash content, chloride content, phosphorous content, Sulphur content, ash constituents, heavy metals, etc. It is desired, therefore, that the AFR mix is designed appropriately by taking into consideration various available waste materials into an appropriate AFR mix having desired properties. Subsequently, the design of the Fuel mix is carried out by taking into consideration the properties of the AFR.

12.3.5   Burner Momentum

Experience has shown that the burner's firing wastes and AFRs require higher momentum than the burner firing normal fossil fuels. It describes the efficiency of the burner to mix the hot secondary air at about 100 °C with the cold (ambient) primary air. The primary air is added to the process at high pressure and velocity (150–250 m/s). There are two kinds of momentum—radial and axial. The burner momentum is a sum of both. Burner momentum is measured as N/MW. It is calculated using the following equation:

$$\text{Momentum (I)} = \frac{(M) \times (V)}{H} \frac{kg}{\sec} \frac{m}{\sec} \frac{1}{MW} = \frac{N}{MW},$$

where M is the Primary air mass flow rate (kg/sec), V is the velocity of air at the burner tip (m / sec) and H is the thermal energy input (MW).

The burner momentum is required to be optimized to an appropriate value to fire AFRs depending upon their burnability characteristics. Typically, a burner momentum of 6–8 N/MW may be good enough for the pulverized coal and the same would be in the range of 9–12 N/MW for some of the AFRs such as plastic waste, Dried Sewage Sludge, other pre-processed AFRs, etc.

12.3.6   Odour Control

We encounter many issues while handling, pre-processing, and co-processing Wastes and AFRs. One of the major issues is odour and it needs to be managed appropriately because it is a big irritation concern to different stakeholders such as employees, community, plant visitors, etc. Major factors relevant to odour nuisance are the following:

- Offensiveness,
- Duration of exposure,
- Frequency of odour occurrence,
- Tolerance and expectation of the receptor.

In many cases, the community tends to forcefully stop the plant activities from which the odour is emanating. Hence, appropriate intervention is required to control the odour.

There are different technologies that can be implemented for odour control (CPCB, 2008).

A.   Odour Control from Area Sources

For large area sources following methods can be used to reduce odour complaints:

i.   Avoiding development of community close to the site and development of green belt in the in-between zone
ii.   Use of atomizers that can spray ultra-fine particles of water or fragrances or chemicals along the boundary lines of the area to suppress odours.

B.   Odour Control from Point Sources

The point source odour causing gas stream can be treated using different technologies such as mist formation, thermal oxidation, catalytic oxidation, biofiltration, adsorption, wet scrubbing, chemical treatment, masking, condensation, etc. In the context of pre-processing of wastes into AFRs and co-processing of waste /AFRs, some of them are more appropriate and are discussed below.

i.   Pre-processing facility located far away from cement plant:
      Following technologies can be implemented in pre-processing facilities:
      Misting Systems:
      High-pressure fog systems can be installed by adding nozzle rings on the pipework that is installed along the inside periphery of the pre-processing facility. Appropriate odour control blends are added to the water supply system and then pressurized into a fine fog. The released fog along with the odour control blend neutralizes the odour in the facility. It is important to ensure that the fog does not produce wetness on equipment, people, or the floor.
      Catalytic or Thermal Oxidation:
      The gases from the pre-processing facility can be exhausted using properly designed ducts and fans and these gases are then sent through an oxidation

system that is catalytic of thermal so that the odour causing gases are completely combusted, thereby treating the odour issue.

Adsorption on Activated Carbon:

The gases from the pre-processing facility exhausted using ducts and fan are sent through an activated carbon-based adsorption system wherein odour causing gases are picked up and clean air is released. The activated carbon adsorbed with odour causing chemicals can then be taken out and sent for co-processing.

Absorption on Zeolite based Mesh:

Mesh that has a coating of zeolites have the capability of absorbing the odour causing chemicals. This mesh can be installed inside the pre-processing facility in the direction of the wind. The mesh picks up the odour causing compounds and reduces the odour issue faced in the facility. After certain time, when their absorption capability reduces, this mesh can then be washed with a spray of water. This water needs to be sent for the effluent treatment or can be co-processed inside the kiln. (Please refer Chap. 10 for more details).

ii. Pre-processing or co-processing facilities located within the cement plant:

The odour control technologies of misting system, catalytic or thermal oxidation, and the technologies of adsorption on activated carbon or zeolite-based mesh mentioned in the earlier section can be implemented at locations where pre-processing or co-processing facility is set up within the cement plant.

Further, another very powerful technology option available in this case is the use of the high temperature available in the kiln system to combust the odour-causing compounds. For this, the exhausted gases from the pre-processing and/or co-processing facility are sent into the hot gas zone of the kiln system where the odour compounds present in it get fully combusted and mitigate the odour issue.

12.3.7 Occupational Health Hazards and Safety Aspects

Waste materials may include both hazardous and non-hazardous materials derived from Municipal, Industrial and Agricultural sectors. AFR pre-processing and co-processing facility handles hundreds of tonnes of many chemicals. They may come at different frequencies, in different packaging sizes, etc. Further, these wastes do not have properly designed Materials Safety Data Sheets (MSDS). AFR is derived from wastes, and hence its every receipt must be assessed based on a standard format for each of the supplies of materials from each of the new and existing vendors.

The safe management of waste/AFR has to be embedded into the system right at the initial stage and health and safety aspects need to be designed into the system at the planning stage itself. Following are some of the important considerations:

A. *Safety information on the waste/AFR materials*

i. Collate relevant safety information by interacting with the waste generator.
ii. Interact with the doctor affiliated to the waste generating agency for any additional relevant information.

iii.   Obtain a broad idea about the manufacturing process in which waste is generated along with major raw materials utilized.
iv.   Try to gather safety-related information about the waste stream, raw materials, and manufacturing process from the literature sources.
v.   Obtain information on the safe practices and PPEs employed by the waste generating agency.
vi.   Material Safety Data Sheet (MSDS) is generally not available for the AFR/waste streams. It is very useful to utilize AFR/Waste profile sheet provided in Annexure 13 of Guidelines published by GIZ/LafargeHolcim to collate the safety data of each of the waste streams (LH-GIZ, 2020).

B.   *Implementing pre-processing and co-processing facility*

i.   While designing equipment, system and facility take into consideration the waste/AFR-related safety data sheet.
ii.   Procure equipment and systems meeting the required safety standards of the wastes/AFRs to be handled/processed in them.
iii.   Install proper Fire detection and control system, Electrical and Instrumentation system, and spill control mechanism.
iv.   Ensure that the floor is made impervious with the proper concreting arrangement and with geopolymer if hazardous and other chemicals are going to be handled in the facility.
v.   Ensure that all statutory and safety provisions are made while implementing the systems and facility.

C.   *Operation and management of the facility*

i.   Carry out a proper risk assessment of handling, storage, and processing of the waste/AFR material in the facility by conducting HAZOP.
ii.   Use appropriate PPEs while operating the facility.
iii.   Train all the manpower at the facility in the safety monitoring and control system.
iv.   Ensure that fire safety drills are conducted regularly in the facility.
v.   Implement regular fire safety audits.
vi.   Prepare and maintain up-to-date the Emergency Response Plan.

D.   *Aspects related to health and Safety of the operating personnel*

The personnel operating in the pre-processing and co-processing facilities need to be made fully aware of the health and safety aspects related to the wastes and they also need to be trained on mitigating concerns associated with them. The data compiled in the AFR/waste profile sheet referred to in the heading "A" above can facilitate this process to desired satisfaction level.

All the personnel operating in the pre-processing and co-processing facilities need to be tested for their fitness for the job as per the applicable statutory guidelines defined by the applicable authorities. This evaluation requires medical investigation of the operating personnel including some of the specifically mandated pathological tests. In India, currently, the same is required to be carried out every six months.

Care is required to be taken to ensure that the contamination of wastes to the clothes of the workers is safely contained and treated. For this, a washing arrangement needs to be set up at the facility and the wash water is required to be sent to the effluent treatment facility.

It is also important that the food and other eatables of the workers are not contaminated with wastes. For this, by design, segregation of the workplace and canteen is ensured and proper hygiene practices mandated.

12.4 Conclusions

Pre-processing and co-processing of wastes raises many operational issues, and they need to be tackled well to be successfully able to achieve the desired level of AFR utilization in cement clinker manufacture. These include concerns related to sustainability, statutory, quality variation in wastes, manpower and their skills, process emergencies, stakeholder engagement, design considerations, receipt and storage of AFRs, handling and feeding systems, receipt of non-compliant materials, quality control, process control, various technical considerations, etc. These concerns need to be mitigated and it can be seen from the above deliberations that there are appropriate technology and other options available to successfully implement co-processing successfully, smoothly, and safely.

References

CII-Godrej GBC. (2010). Cement Formulae Handbook published by CII-Sohrabji Godrej Green Business Centre.

CPCB. (2008). Guidelines on odour pollution & control published by CPCB, May 2008. Retrieved from http://www.espair.co.in/download/odor/cpcb-guideline-odor-management.pdf

LafargeHolcim/GIZ. (2020). Guidelines on Pre & Co-processing of wastes in Cement production, Annexure 13, pp. 119–122.

Lowes, T., & de Souza, J. B. (2017). Preventing Build-ups and rings. *International Cement Review,* 78–83.

Montes de Oca, P., & Forinton, J. (2010). Preheater Blockages Event: Ficem-APCAC Technical Conference, Montego Bay, Jamaica. Retrieved September 08 2010, https://www.ficem.org/mul timedia/2010/tecnico10/19_Jonathan_Forinton_ATEC.pdf

Part VI
Co-processing: Business Models, Case Studies, Global Scenario, Growth and Advocacy

Chapter 13
Case Studies and Business Models in Pre and Co-processing

13.1 Introduction

Utilization of different kinds of wastes as Alternative Fuels and Raw Materials (AFR) is implemented successfully by many cement plants in many different countries. This practice is in operation for more than the last four decades. In almost all the cement plants, it is practiced as a business initiative. The major business principle employed by the cement plants while undertaking co-processing is that the use of wastes as AFRs must reduce their cost of production. This reduction in the production cost of clinker is the driver of this waste management business initiative. The production cost can be reduced by reducing the variable cost of production. There are different business models that get employed while implementing co-processing in cement kilns based on several appropriate parameters. It was very interesting to note that for many years, some of the plants in Europe had a negative variable cost of production. The various economic considerations put forward in the following sections will illustrate the salient features of the AFR business models in detail.

13.2 Economic Parameters Utilized in Co-processing Business

For ensuring a sustainable business model, the project feasibility needs to be looked into. There are a number of parameters that need to be assessed to develop effective and sustainable business models. This section will discuss the economic parameters. Such as: Production Cost of Clinker manufacture which is broken down into several related parameters, Transport costs, Gate/Tipping Fee, Price of AFR, Savings from the use of AFR, Cost of the Pre-processing facility, and Cost of the Co-processing facility. These parameters will be described in the following subsections.

© The Author(s), under exclusive license to Springer Nature Singapore Pte Ltd. 2022
S. K. Ghosh et al., *Sustainable Management of Wastes Through Co-processing*,
https://doi.org/10.1007/978-981-16-6073-3_13

13.2.1   Production Cost of Clinker Manufacture

This cost is the sum of variable cost and the fixed costs incurred in the production of clinker.

A.   Variable cost of Clinker Production:

Variable costs are linked to the volume of production and varies accordingly. Lower the production volume lower are the variable costs and vice versa. In clinker production, these refer to the costs of raw materials, fuels, electricity utilized, and transport cost incurred for sending the material to the market etc.

B.   Fixed costs in Clinker Production:

Fixed costs are independent of the volume of production and remain the same irrespective of the volume of production. In clinker production, these refer to the costs of manpower, rent, interest, depreciation, advertisement, other overheads, etc.

III.   Raw material Costs:

These relate to the procurement price of raw materials such as Limestone, Iron ore, Bauxite, Clays, etc. These are procured based on the price per tonne of the relevant constituent. For example, if the Iron content in Iron ore is 60%, then while determining the price of iron ore, this 60% iron content is taken into consideration.

IV.   Fuel costs:

Fuel costs are related to the energy content present in the fuel. For example, if the calorific value of one fuel is 5000 Kcal/Kg and the same is priced at USD 70/T, then the fuel having a lower Calorific value than 5000 Kcal/Kg will get sold at lower price and fuel having a calorific value higher than 5000 Kcal/Kg will attract higher price.

E.   Electricity costs:

These relate to the cost of electricity utilized in clinker production. For example, if 25 KWh of electricity is utilized for the production of 1 T of clinker, then the electricity cost will be the purchase cost of 25 KWh of electricity.

13.2.2   Transport Costs

These relate to the distance of travel, waiting time, loading–unloading times, etc. For example, if the transport distance is 100 km, then the transport cost will be accounted based on 100 km of travel of the vehicle, waiting time and the time & effort required to load, unload, etc. The costing for the transport gets worked out on USD/T/Km basis taking into consideration all the costs as mentioned.

13.2.3 Gate/Tipping Fee

Gate fee or Tipping fee is the amount waste generator pays to the agency that manages its waste as per the acceptable norm. This is a service fee charged by the waste management agency to the waste generator. This is represented as USD/T of the waste.

13.2.4 Price of AFR

It is the amount that the agency that owns the AFR, charges to the agency that wants to purchase it. It is represented in USD/T of AFR. If it is an Alternative Fuel, it will get linked to the calorific value and if it is an alternative raw material, then it gets linked to the specific raw material content in the AFR.

13.2.5 Savings from Use of AFR

It is the savings benefit that the cement plant will be deriving by replacing natural fuels and raw materials with AFR. It is represented as USD/T or USD/1000 kcal.

13.2.6 Cost of the Pre-processing Facility

The cost of the fully constructed pre-processing facility includes the costs of land, civil and structural works, process equipment and machinery, laboratory, material handling equipment, firefighting, and environmental control facilities, etc. It is represented in USD.

13.2.7 Cost of the Co-processing Facility

The cost of the fully constructed co-processing facility includes the costs of land, civil and structural works, process equipment and machinery, laboratory, material handling equipment, firefighting, and environmental control facilities, etc. It is represented in USD.

13.3   Principles of AFR Business

The AFR Business consists of the following considerations.

Waste materials do not have any defined quality or market acceptability. Hence, they do not get sold in the marketplace. They need to be disposed or sustainably managed as per the direction given under the rules. Both options require money to be spent.

Following are typical examples of some of such wastes:

- ETP sludges from chemical or other industries.
- Distillation residues from chemical and other industries.
- Date expired medicines, pesticides, banned products, etc.
- Date expired FMCG products, Chocolates, Juices, food items, seeds, etc.

These wastes used to be sent for landfill or incineration earlier for disposal and now they are sent to cement kiln for co-processing for managing them in a sustainable manner. Before sending them for co-processing, they need to be processed into resources which are called as AFRs. This operation is called as pre-processing of wastes into AFRs.

Some waste streams have uniform quality such as Rice husk and other biomasses and they can be utilized as resources (AFR) as such in cement kilns. They do not need any specific processing. Wastes of this kind attract price in the market for utilization.

The difference between wastes and resources is clearly visible in Fig. 13.1.

1.Resource

Co-processing of AFRs in cement kiln requires setting up of appropriate facilities for pre-processing and co-processing which calls for investment and operational

Fig. 13.1 Illustration of wastes and resource

cost. Hence, AFR co-processing must be based on an appropriate business model that provides a decent investment return to the cement plant.

13.4 Concepts Related to Costs, Prices and Viability of Use of AFRs

Waste is a liability, and hence its impact on environment, safety requirements, and economy of use needs to be evaluated. To assess the impact on the environment and safety aspects, the cost implications associated with the required protection measures need to be analyzed. To understand the impact on the economy of use, the likely monetary benefit that they will bring to the cement plant needs to be evaluated.

Following are the various concepts related to price cost, and viability of the use of AFRs:

13.4.1 Cost of Natural Raw Material

These costs are the used costs of the raw materials in the cement manufacture.

13.4.2 Cost of Natural Fuel

These costs are the fired cost of the natural fuel in the cement manufacture.

13.4.3 Operation and Management Cost

This includes the cost towards electricity, manpower, diesel for material handling equipment, safety and environment control, testing and quality control, maintenance costs of equipment and facility, including interest and depreciation, etc.

A.   Pre-processing cost:

This cost pertains to the operation and management of pre-processing of wastes into AFRs. In case this pre-processing is carried out in the cement plant, then the relevant cost towards this activity also needs to be accounted for in the viability assessment of the use of AFRs in the cement plant.

B.   Co-processing Cost:

This cost pertains to the operation and management of co-processing facility through which the AFRs are fed into the cement kiln via calciner or kiln inlet or main burner or mid kiln option.

13.4.4   Resource Replacement Cost

AFR may replace fuel or raw material and hence it is called as Alternative Fuel (AF) or Alternative Raw material (AR). The contribution that AF or AR will make in the cement plant economics depends upon the resource replacement cost. As an example, in cement manufacture, AFR may replace coal or AFR may replace raw material. This replacement cost needs to be calculated suitably.

Typical costing examples of co-processing AFR as replacement of coal and replacement of Iron ore are illustrated in Tables 13.1 and 13.2, respectively. The cost of the fuel is always represented as USD per 1000 kcal or USD Per Kilo Joules and the cost of raw material as USD Per Tonne.

I.   AFR replacing Coal

In Table 13.1, a typical example of estimating the maximum price of AFR at which it can substitute coal is explained.

Table 13.1 Typical example—AFR replacing coal

S. No	Parameter	Value
1	As fired cost of Coal	USD 0.014 per 1000 kcal
2	Maximum payment that cement plant can make for AFR	USD 0.014 per 1000 kcal
3	NCV of AFR	3000 (Kcal / Kg)
4	Maximum price that cement plant can pay for the ready to fire AFRs	USD 42 / T

The actual payment that the cement plant would be making for the Alternative Fuels will be lower than this number. This is because it has to take into account the operational cost of co-processing, interest, and depreciation charges of the facility, impact of moisture, ash and burnability of the AFR on production, etc.

Table 13.2 Typical example—AFR replacing Iron ore

S. No	Parameter	Value
1	Iron content in Iron ore	60%
2	As used cost of Iron ore	USD 14 / T Iron ore
3	Cost of Iron (=14 / 0.6)	USD 23.2 / T Iron
4	Iron content in AFR	75%
5	Maximum price feasible for AFR (=23.2*0.75)	USD 17.4 / T AFR

In the case of Alternative Raw materials also, the actual payment that the cement plant would be lower than this number

II.   AFR replacing Iron Ore

In Table 13.2, a typical example of estimating the maximum price of AFR is explained at which it can substitute existing natural Iron source.

III.   Treatment cost of hazardous or non-hazardous wastes by landfill or incineration options

Management of Industrial hazardous and non-hazardous wastes needs to be carried out as per the processes defined in the rules and regulations. Prior to the hazardous and other waste management Rules, 2016, in India, the prescribed options were landfill and incineration. In the HOWM Rules, 2016 notified by Government of India, co-processing was incorporated as an option for the management of hazardous and non-hazardous wastes.

The cost incurred by the waste generator for the management of waste through the alternative treatment option of landfill or incineration is an important parameter for defining the costs related to the co-processing option. The waste generator would be more inclined to opt for co-processing if the cost of co-processing is cheaper than the other alternative options of landfill and incineration options.

This cost concept is illustrated for the incineration and landfill options in Tables 13.3 and 13.4, respectively.

IV.   Delivered cost of biomass/agro-waste at the cement plant

The biomass is generated in the grain processing or food processing plants and the agro-waste gets generated during farming or agricultural activities. Biomass such as rice husk, Bagasse, sawdust, etc., is fairly uniform in quality and can be fed without any additional processing, and hence its price comparison tends to happen in the marketplace with that of the fossil fuel. There would be some cost implications on account of moisture content, requirement of shredding, etc., which will have to be considered by the cement plant.

Table 13.5 illustrates the price of Biomass that gets assessed as acceptable at the cement plant.

Table 13.3 Typical example—Treatment cost for the option of Incineration

S. No	Parameter	Value
1	Name of the waste requiring incineration option (CV = > 2500 kcal /Kg)	Distillation Residue
2	Existing treatment option for disposal	Incineration
3	Existing treatment cost of disposal	USD 208 / T waste

Hence, if the co-processing cost is lower than USD 208/T of Distillation Residue, then the waste generator would be having an incentive to opt for co-processing

Table 13.4 Typical example—Treatment cost for the option of landfill

S. No	Parameter	Value
1	Waste generated by Industry (CV = < 2500 Cal / gm)	ETP Sludge
2	Existing treatment option for disposal	Secured Landfill
3	Secured Landfill cost	USD 42 / T of waste

Hence, if the co-processing cost is lower than USD 42/T of ETP sludge, then the waste generator would be having an incentive to opt for co-processing. Co-processing of the waste helps the waste generator to achieve zero-landfill status. This also acts as an incentive for the waste generator

Table 13.5 Typical example—Sourcing of Biomass

S. No	Parameter	Value
1	Cost of coal (From Table 13.1)	USD 0.014/1000 kcal
2	Maximum payment that cement plant can make for receiving Biomass at its plant	USD 0.014/1000 kcal
3	Net Calorific Value of Biomass	3000 kcal/kg
4	Maximum acceptable Price of Biomass delivered at cement plant	USD 42/T Biomass

The actual price paid by the cement plant would be lower and depend upon the cost implications on account of shredding required, moisture removal, etc

V.   Delivered cost of SCF/RDF at the cement plant

SCF is an un-processed Segregated Combustible Fraction derived out of dry MSW. It has no uniform quality or size fraction. Its uniform Net Calorific Value (NCV) would be < 2000 kcal/Kg. RDF is a processed AFR. It is derived out of a dry combustible fraction of MSW meeting uniform quality and standard size acceptable by the cement plant. Generally, RDF has NCV of less than 4500 kcal/Kg. SCF does not have any size specifications. RDF has a size specification of < 75 mm or < 50 mm or < 30 mm, etc.

Table 13.6 Typical example—Sourcing of SCF / RDF

S. No	Parameter	Value
1	Cost of coal (From Table 13.1)	USD 0.014/1000 kcal
	Maximum price that cement plant will make for receiving SCF or RDF at its plant (From Table 13.5)	USD 0.014/1000 kcal
2	Maximum average NCV of SCF	2000 kcal/Kg
4	Maximum price at which SCF would be sourced by cement plant	USD 28/T SCF[a]
5	Max NCV of processed RDF	4500 kcal/kg
6	Maximum price at which RDF would be sourced by cement plant	USD 62.5/T RDF[b]

[a] The price of SCF workable to cement plant would be in the range of 25% to 40% of this price. This reduction level in price depends upon the cost that the cement plant will have to incur to improve its quality to the desired level and also the impact that moisture, ash, Chlorine content, and burnability of SCF will have on the clinker process and production.
[b] The price of RDF would be about 50% to 70% of this cost. This reduction level in price depends upon the cost that the cement plant will have to incur to improve its quality to the desired level and also the impact that moisture, ash, Chlorine content, and burnability of RDF will have on the clinker process and production.

Typically, smaller size (say shredded to < 30 mm) SCF or RDF attract higher price than the larger size materials depending upon their NCV.

Table 13.6 illustrates the typical maximum pricing of SCF/RDF at which it will get received at the cement plant.

VI.   Viability of using AFR in the cement plant

AFR is utilized in the cement plant for replacing the fossil Raw materials and Fuels that are extracted from nature by mining. These include fuels such as coal, lignite, oil, gas, etc., and raw materials such as limestone, iron ore, bauxite, clay, etc. While the use of AFRs will be replacing the natural materials, the same will be workable only when the cost of AFR is lesser than that of natural materials. Further, to process wastes into AFR and feed AFR into the cement kiln, additional infrastructure is required at the cement plant which will be requiring capital expenditure to install it and operating expenditure to run it.

The saving that the cement plant will be making by using AFR at the plant will have to provide adequate returns on the investment made by the cement plant. If these returns are acceptable to the cement plant, then the use of the AFR is considered as viable and the business model is considered as acceptable. Each organization will have its own methodology and parameters for this kind of evaluation and acceptance.

The basic parameters in such viability calculations are the price of the materials, the cost of operations and the investments made, and payback period or return on investment are the guiding factors. Usually, a payback period of less than 5 years is considered as a viable proposition. Of course, there could be some additional considerations that cement plant will have to take into consideration depending upon the government mandates or organizational policies.

The various costs/prices considered in the analysis in this chapter are typically Indian and prevailing as of 2021.

13.5   Business Models with Pre-processing and Co-processing

Co-processing requires feeding the AFR materials in the cement kiln through calciner, kiln inlet, or main burner. Depending upon the capacity of co-processing, an elementary manual system or a standard mechanized system is implemented. Generally, the manual arrangement is ok for co-processing <2 TPH of AFR, and a mechanized system would be required for capacity >2 TPD. The pre-processing facility is generally required when the desired TSR% is >3% depending upon the quality variation in the waste streams.

Following Six types of facilities are considered for illustrating the different business models:

1.   Elementary Co-processing Facility for Solid AFRs,
2.   Elementary Pre-processing & Co-processing Facility for Solid AFRs,
3.   Mechanized co-processing Facility for solid AFRs,
4.   Mechanized Pre-processing Facility for Solid AFRs,
5.   Co-processing facility for liquid AFRs,
6.   Co-processing of Alternative Raw material through Raw meal.

The different business models are described below. In the analysis of the business model assessment, viability based on the price of materials and costs of operation is considered for illustration purposes. Detailed evaluations can be made by the organizations based on the prevailing organizational practices and governmental statutes.

13.5.1   Business Model—Elementary Co-processing Facility for Solid AFRs

The co-processing facility consists of a covered shed to store the received processed AFR and a payloader to handle and transport it. This processed AFR is then conveyed up to the calciner floor using a winch-based system having a capacity of about 2 TPH. The AFR is fed into the calciner using a double flap valve and shut gate assembly. A laboratory facility is created to analyze the properties of the received AFR. Basic firefighting facility is also considered as implemented.

The typical costs of such a facility along with the various items present in it are provided in Table 13.7.

Table 13.7 Typical cost of an elementary co-processing facility

S. No	Item	Capacity	Cost in USD
1	Double flap valve & shut gate	200 M3/Hr	2,80,000
2	Pay loader	~5 TPH	42,000
3	Storage shed	<300 M2	70,000
4	Winch based conveyor	~2 TPH	42,000
5	E&I & Firefighting	As required	70,000
6	Laboratory	As required	70,000
7	Structural work	As required	42,000
	Total		6,16,000

I.   Example—Co-processing of Biomass:

Biomasses are excellent alternative fuels for co-processing in cement kilns. Being reasonably uniform in quality, safe to use, and easy to procure, these AFRs are utilized by many agencies that utilize fossil fuels such as cement plants, power plants, brick kilns, steam boilers, etc.

These materials, being in demand, get sold at a decent price in the marketplace. Cement plants can directly feed them in the kiln through calciner. They can also be co-processed through the main burner if they are shredded finer in size. The price of these materials is close to that of the fossil fuel and the viability of their use depends upon the availability in the surrounding region because the transportation cost has to be reasonably less so as to make the delivered price acceptable at the cement plant.

Biomasses are carbon neutral, and hence are very useful in reducing the carbon footprint of the cement plant. The best option for the cement plant is to create a cooperative society of the farmers located within a reasonable distance of the plant and obtain the biomass produced by them at a price workable for both.

II.   Business Model of co-processing:

Figure 13.2 depicts the business model of co-processing of biomass.

III.   Salient Features of the Business Model:

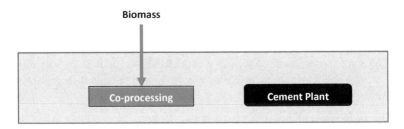

Fig. 13.2 Business model for Co-processing of Biomass/Agro-wastes

Table 13.8 Salient Features of business model for co-processing of Biomass

S. No	Parameter	Value
1	Investment in co-processing facility (Table 13.7)	USD 616,000
2	Cost of coal (From Table 13.1)	USD 0.014 / 1000 kcal
3	Negotiated price of biomass (Assumed)	USD 27 / T
5	Co-processing cost (Typical)	USD 0.28 / T
6	Total cost	USD 29.9 / T
8	NCV of Biomass	3000 kcal / Kg
4	Total cost of biomass co-processing	USD 0.01 / 1000 kcal
7	Savings benefit of using biomass (0.014–0.01)	USD 0.004 / 1000 kcal
9	Biomass used per day	40 TPD
10	Biomass used per year (300 day / year)	12,000 Tons per year
11	Total energy from biomass	36,000,000,000 kcal
12	Savings benefit derived by using Biomass	USD 142,800 per year
13	Payback Period of investment	4.31 years

Table 13.8 provides the salient features of this business model with typical values for the co-processing of biomass using the elementary co-processing facility.

13.5.2 Business Model—Elementary Pre-processing and Co-processing Facility for Solid AFRs

This pre-processing facility consists of a shredder installed in a storage shed. This shed stores the incoming and processed AFR. A pay loader is used for handling and transport of incoming and processed materials. A laboratory facility is created to prepare the solid sample & to evaluate its basic properties. A basic firefighting facility is also considered as installed in the facility.

The typical costs of the various items in such a facility are provided in Table 13.9

I.   Example—Co-processing of RDF prepared from SCF:

The Segregated Combustible Fraction (SCF) from MSW is available from the following sources:

i.   Integrated MSW management sites of the towns and cities.
ii.   Dry waste segregation taking place at the MRFs set up in the towns and cities.
iii. Dump-yard remediation activities are being implemented in the towns and cities.

These different sources are depicted in Fig. 13.3.

It is desired that the SCF generating cities and towns are in the close vicinity of the cement plant so as to receive the same at a reasonably lesser cost.

Table 13.9 Typical cost of an Elementary Pre-processing and Co-processing facility

S. No	Item	Capacity	Cost in USD
1	Double flap valve & shut gate	200 M3/Hr	280,000
2	Shredder (Indian)	~5 TPH	84,000
3	Pay loader / forklift	~5 TPH	42,000
4	Storage shed	<300 M2	70,000
5	Winch based conveyor	~2 TPH	42,000
6	E&I & Firefighting	As required	98,000
7	Laboratory	As required	98,000
8	Structural work	As required	42,000
9	Total		756,000

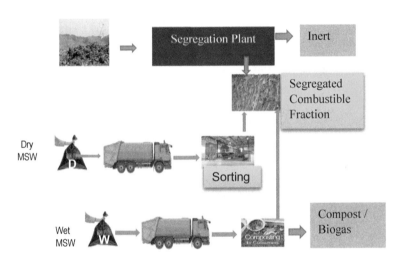

Fig. 13.3 SCF generation in towns and cities

As SCF is a mix of many materials, it does not have desired uniform characteristics for use as Alternative Fuel in the cement plant. SCF is, therefore, required to be pre-processed in a pre-processing facility into RDF before sending the same to the kiln for co-processing. In the pre-processing facility, the material is blended to achieve desired quality consideration and is shredded then to the desired size fraction. The pre-processing of SCF to RDF of desired quality can be set up by a third-party agency or the same can be done by a cement plant as well.

II.   Business Model of Elementary Pre-processing and co-processing:

The business model for use of SCF/RDF in cement kilns is depicted in Fig. 13.4.

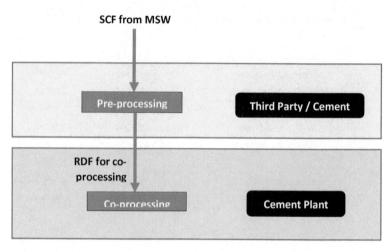

Fig. 13.4 Business model for Co-processing of SCF / RDF from MSW

III.   Salient Features of the Business Model:

Table 13.10 provides the salient features of this business model with typical values for the co-processing of RDF prepared from SCF in a facility having pre-processing and co-processing facilities.

Table 13.10   Salient features of the business model for pre-processing SCF/Co-processing of RDF

S. No	Parameter	Value
1	Investment in pre & co-processing facility (Table 13.9)	USD 756,000
2	Cost of coal (From Table 13.1)	USD 0.014 / 1000 kcal
3	Negotiated price of SCF (Assumed)	USD 10.5 / T SCF
4	Yield of RDF from SCF	60%
5	Price of SCF in terms of RDF (=10.5 / 0.6)	USD 17.5 / T RDF
6	Pre-processing cost	USD 4.2 / T RDF
7	Co-processing cost (Typical)	USD 2.8 / T RDF
8	Total cost	USD 24.5 / T RDF
9	NCV of RDF	3000 kcal / Kg
10	Total cost of RDF co-processing	USD 0.008 / 1000 kcal
11	Savings benefit of using RDF (0.014–0.008)	USD 0.006 / 1000 kcal
12	RDF used per day	30 TPD
13	RDF used per year (300 day / year)	9000 Tons per year
14	Total energy from RDF	27,000,000,000 kcal
15	Savings benefit derived by using RDF	USD 162,000 per year
16	Payback Period of investment	4.8 years

13.5.3 Business Model—Mechanized Co-processing Facility

The mechanized co-processing facility consists of the shed to store the received pre-processed AFR and a payloader to handle and transport it. This pre-processed AFR is then conveyed up to the calciner floor using a long belt conveyor. Dosing to the conveyor is done by using walking floor and weigh feeder. The system has an operating capacity of about > 20 TPH. The AFR is fed into the calciner using a double flap valve and shut gate valve. A laboratory facility is created to analyze the properties of the received AFRs in detail. An appropriate firefighting facility is also implemented.

The typical costs of the various items in such a facility are provided in Table 13.11.

I.   Example—Co-processing of a Mix of different kinds of Solid Processed AFRs:

It is possible to obtain the pre-processed AFRs of different kinds and blend them at the cement plant and co-process them. These include pre-processed hazardous solid wastes, pre-processed non-hazardous solid wastes, tyre chips, RDF meeting desired physico-chemical characteristics, and biomass.

The industrial hazardous wastes consist of distillation residue, ETP sludges, failed batches, chemical sludges, used catalysts, etc. Industrial non-hazardous wastes consist of expired FMCG, food and beverage products, date expired seeds, ETP from water treatment plants, etc. These materials need to be disposed of by incineration or landfill options which cost a reasonable money. If co-processing is feasible at prices cheaper than the same, then it is a win–win solution for both wastes generating industry and the cement plants.

Here the cement plant gets paid for the service extended by him for managing the waste sustainably in the cement kiln. This service fee depends upon several

Table 13.11 Typical cost of a Mechanized Co-processing facility

S. No	Item	Capacity	Cost in USD
1	Double flap valve & shut gate	200 M3/Hr	280,000
2	Pay loader (2 nos)	>20 TPH	105,000
3	Storage shed	<2000 M2	420,000
4	Long belt conveyor (>250 M)	>20 TPH	350,000
5	Walking floor	>20 TPH	280,000
6	Weigh Feeder	>20 TPH	28,000
7	Short belt conveyor (<8 M)	>20 TPH	7,000
5	E&I & firefighting	As required	140,000
6	Laboratory	As required	98,000
7	Structural work	As required	210,000
	Total		1,918,000

factors such as the constituents present in the waste material, its storage, handling and processing complexity, etc. But for the business model to work, it needs to be lower than the existing cost of disposal incurred by the waste generator. These waste streams are pre-processed into uniform materials by the waste management agencies or the same can be done by the cement plant as well.

The tyre chips are available in many countries derived out of used tyres. They may be available in the country or may be imported. The cost depends upon several factors. In some countries, the tyre chips get paid or delivered free at the cement plant, and in some other countries they may be available for a price like any other waste materials.

The SCF is generally available from the material recovery facilities or integrated MSW management facilities or dump yard remedying processes. These need to be properly segregated by removing the inert and recyclable materials and then treated by shredding and blending process to achieve RDF having desired size and fuel properties. This processing of SCF into RDF can be done by third parties or the cement plant. The price of RDF depends upon the price of coal at the cement plant.

Biomass can be obtained from the grain or food processing industries or from agricultural activity. This needs to be shredded to a certain acceptable size before using it in the cement plant. This processing can be done by a third-party agency or can be done by the cement plant. The price of the shredded biomass depends upon the cost of coal at the cement plant.

In this business model, the pre-processing of hazardous wastes, Non-hazardous wastes, shredding of tyres, Processing of SCF into RDF is carried out by third parties and delivered at the cement plants. This is then blended in the cement plant in a certain ratio and sent to the calciner for co-processing using walking floor and long belt conveyor.

II.   Business Model of Mechanized Co-processing Facility:

The business model co-processing mix of processed AFRs is depicted in Fig. 13.5.

III.   Salient Features of the Business Model:

Table 13.12 provides the salient features of the business model of the mechanized co-processing facility in which a mix of different AFRs pre-processed by different agencies is co-processed.

Table 13.13 provides the calculation of the average NCV and price of the AFR mix consisting of different waste streams.

Table 13.14 provides the salient features of this business model with typical values for the co-processing of AFR mix.

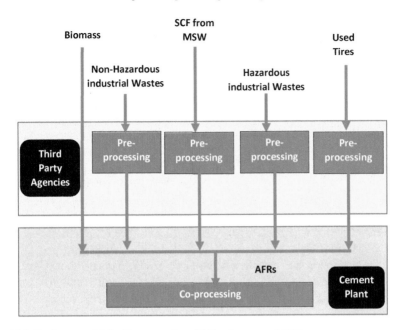

Fig. 13.5 Business model for Co-processing of Mix of processed AFRs

Table 13.12 Data pertaining to constituents in AFR mix

S. No	Parameter	Value
1	Negotiated price of RDF (Assumed)	USD 29.4 / T RDF
2	NCV of RDF	3000 kcal / Kg
3	Quantity of RDF co-processed	80 TPD
4	Negotiated price of Biomass (Assumed)	USD 33.6 / T Biomass
5	NCV of Biomass	3000 kcal / Kg
6	Quantity of Biomass co-processed	60 TPD
7	Price of tyres chips (Assumed)	USD 84 / T Tyre Chips
8	NCV of Tyre chips	7500 kcal / Kg
9	Quantity of Tyre chips co-processed	20 TPD
10	Negotiated Price of Processed Non- hazardous waste	(-ve) 7 / T NHW
11	NCV of NHW	2000 kcal / Kg
12	Quantity of NHW co-processed	30 TPD
13	Negotiated Price of Processed hazardous waste	(-ve) 14 / T NHW
14	NCV of HW	4000 kcal / Kg
15	Quantity of HW co-processed	10 TPD
16	Average price of blended AFR	USD 28.49 / T AFR
17	Average NCV of AFR	3350 kcal / Kg
18	Total AFR co-processed	200 TPD

Table 13.13 Calculation of weighted average Price and NCV of AFR mix

AFR	TPD	% Mix (%)	NCV	Wtd. Av. NCV	USD/T	Wtd. Av. Price
RDF	80	40.0	3000	1200	29.4	11.76
Biomass	60	30.0	3000	900	33.6	10.08
Tyre Chips	20	10.0	7500	750	84	8.4
Processed NHW	30	15.0	2000	300	−7	−1.05
Processed HW	10	5.0	4000	200	−14	−0.7
Total	200	100.0	Average NCV	3350	Average Price	28.49

Table 13.14 Salient Features of the business model for Co-processing of AFR Mix

S. No	Parameter	Value
1	Investment in co-processing facility (Table 13.11)	USD 1,918,000
2	Cost of coal (From Table 13.1)	USD 0.014 / 1000 kcal
3	Negotiated price of AFR Mix (From Table 13.16)	USD 28.49 / T
5	Co-processing cost (Typical)	USD 4.2 / T
6	Total cost	USD 32.69 / T
8	NCV of AFR Mix	3350 kcal / Kg
4	Total cost of AFR Mix co-processing	USD 0.0.0098 / 1000 kcal
7	Savings benefit of using AFR mix (0.014–0,0098)	USD 0.0042 / 1000 kcal
9	AFR Mix used per day	200 TPD
10	AFR Mix used per year (300 day / year)	60,000 Tons per year
11	Total energy from AFR Mix	201,00,00,00,000 kcal
12	Savings benefit derived by using AFR Mix	USD 852,600 per year
13	Payback Period of investment	2.25 years

13.5.4 Business Model—Mechanized Pre-processing and Co-processing Facility

The mechanized pre-processing facility consists of a heavy-duty shredder installed in a storage shed. It also has pits to impregnate hazardous pasty wastes. The impregnation is carried out by blending the same with materials like sawdust. This shed stores the incoming waste materials and the processed AFR. A payloader is used for handling and transport of incoming and processed materials. The mechanized co-processing facility consists of conveying the pre-processed AFR up to the calciner floor using a long belt conveyor. Dosing to the conveyor is done by using walking floor and weigh feeder. The system has an operating capacity of > 20 TPH. The processed AFR is fed into the calciner using a double flap valve and shut gate valve. A laboratory facility is created to analyze the properties of the received wastes and the processed AFRs in detail. An appropriate firefighting facility is also implemented. The typical costs of the various items in such a facility are provided in Table 13.15.

Table 13.15 Typical cost of a Mechanized Pre-processing and Co-processing facility

S. No	Item	Capacity	Cost in USD
1	Shredder (European)	>20 TPH	840,000
2	Pay loader (2 nos)	>20 TPH	105,000
3	Storage shed	<2000 M2	420,000
4	Double flap valve & shut gate	200 M3/Hr	280,000
5	Long belt conveyor (>250 M)	>20 TPH	350,000
6	Walking floor	>20 TPH	280,000
7	Weigh Feeder	>20 TPH	28,000
8	Short belt conveyor (<8 M)	>20 TPH	7,000
9	E&I & Firefighting	As required	350,000
10	Laboratory	As required	210,000
11	Structural Work	As required	210,000
12	Total		3,080,000

Table 13.16 Data pertaining to constituents in the waste mix

S. No	Parameter	Value
1	Negotiated price of Pasty waste (Assumed)	-USD 42 / T
2	NCV of Pasty waste	5000 kcal / Kg
3	Quantity of pasty waste used	40 TPD
4	Negotiated price of Saw dust (Assumed)	USD 63 / T
5	NCV of saw dust	3000 kcal / Kg
6	Quantity of saw dust used	40 TPD
7	Price of Incinerable waste (Assumed)	-USD 42 / T
8	NCV of Tyre chips	2600 kcal / Kg
9	Quantity of Incinerable cost used	40 TPD
10	Negotiated Price of landfillable waste 1 (Gate Fee)	(-ve) 14 / T
11	NCV of landfill waste 1	500 kcal / Kg
12	Quantity of landfilled waste 1 used	40 TPD
13	Negotiated Price of landfillable waste 2 (Gate Fee)	(-ve) 14 / T
14	NCV of landfillable waste 2	800 kcal / Kg
15	Quantity of landfillable waste 2 used (Gate Fee)	40 TPD
16	Average price of AFR Mix	(-ve) USD 9.8 / T AFR
17	Average NCV of AFR Mix	2380 kcal / Kg
18	Total AFR co-processed	200 TPD

I.   Example—Pre-processing of hazardous Incinerable and landfillable Mix of Solid wastes into AFRs and Co-processing the same.

In this example, one kind of incinerable solid waste, one kind of incinerable pasty waste, and two kinds of landfillable hazardous wastes are considered as an example. The pasty waste is impregnated into a free-flowing solid powder by mixing it with sawdust. This impregnated solid waste is then fed into a shredder along with the other incinerable solid waste and the two landfillable solid wastes. The shredded and processed solid AFR is then fed into the walking floor using pay loader. Walking floor feeds the same into the long belt conveyor in a metered manner using weigh feeder. This metered quantum of AFR then gets conveyed to the pre-calciner floor using a long belt conveyor. The processed AFR is then fed into the calciner using a short belt conveyor, double flap valve assembly.

II.   Business Model of Mechanized Pre-processing and Co-processing Facility:

The business model of this case is provided in Fig. 13.6.

III.   Salient Features of the Business Model:

Table 13.16 provides the salient features of the business model of the mechanized pre and co-processing facility in which a mix of incinerable and landfillable wastes is co-processed.

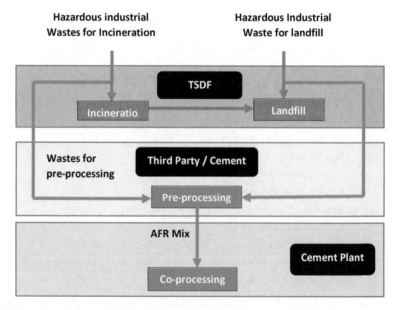

Fig. 13.6 Business model for Co-processing of Hazardous industrial wastes

Table 13.17 provides the calculation of the average NCV and price of the AFR mix consisting of incinerable and landfillable waste streams.

Table 13.18 provides the salient features of this business model with typical values for pre and co-processing of AFR mix of different incinerable and landfillable waste streams.

Table 13.17 Calculation of weighted average Price and NCV of AFR mix

AFR	TPD	% Mix (%)	NCV	Wtd Av. NCV	USD/T	Wtd Av. Price
Pasty waste	40	20.0	5000	1000	−42	−8.4
Saw dust	40	20.0	3000	600	63	12.6
Incinerable solid	40	20.0	2600	520	−42	−8.4
Landfillable solid 1	40	20.0	500	100	−14	−2.8
Landfillable solid 2	40	20.0	800	160	−14	−2.8
Total	200	100.0	Average NCV	2380	Average Price	−9.8

Table 13.18 Salient features of the business model for Co-processing of AFR Mix

S. No	Parameter	Value
1	Investment in pre and co-processing facility (Table 13.15)	USD 3,080,000
2	Cost of coal (From Table 13.1)	USD 0.014 / 1000 kcal
3	Negotiated price of AFR Mix (From Table 13.16)	-USD 9.8 / T
5	Pre-processing & Co-processing cost (Typical)	USD 7.0 / T
6	Total cost	-USD 2.8 / T
8	NCV of AFR Mix	2380 kcal / Kg
4	Total cost of AFR Mix co-processing	-USD 0.00118 / 1000 kcal
7	Savings benefit of using AFR (0.014- (-0.0018)	USD 0.01518 / 1000 kcal
9	AFR Mix used per day	200 TPD
10	AFR Mix used per year (300 day / year)	60,000 Tons per year
11	Total energy from AFR Mix	142,80,00,00,000 kcal
12	Savings benefit derived by using AFR mix	USD 2,167,200 per year
13	Payback Period of investment	1.42 years

13.5.5 Co-processing Facility for Liquid AFRs

In this facility, liquid AFR is received in tankers and stored in a tank. The facility has unloading pumps, storage tank, and AFR feeding pumps and pipelines with required filters and valves, etc. The feeding of AFR to the calciner happens with a pipeline and nozzle. The facility has a basic quality control facility as well as firefighting facility.

The typical costs of the various items in such a facility are provided in Table 13.19.

The investment made in the pre-processing and co-processing facilities is required to be paid back based on the revenue earned by the cement plant using Alternative Fuels and Raw materials. This revenue amount includes the savings derived out of the replacement of the natural resources with AFRs and the net revenue earned as the co-processing fees for managing the industrial hazardous and non-hazardous wastes.

I.   Example—Processed liquid AFR from third-party agency.

In the industry, different kinds of liquid hazardous wastes get generated. These include organic wastes that have high calorific value and aqueous waste which have high moisture content. The industries that generate these waste streams include chemical industries, pharmaceutical industries, petrochemical industries, etc. Both these streams can be co-processed in the cement kiln. These liquid waste streams may contain highly toxic substances and it is important that they are properly pre-qualified through laboratory evaluations. These waste streams being incinerable are very expensive to dispose of, and hence they are generally provided for managing in a sustainable manner with a gate fee.

Liquid waste co-processing requires a separate feeding system and depending upon the chemical nature of the material, they may be fired in the main burner or the calciner.

The feeding system consists of unloading pumps, storage tank, filters to remove coarse and fine impurities, feeding pumps, piping and nozzle to fire through the main burner or calciner. These facilities may require flameproof installations and higher level of fire protection system if the flashpoints of these waste streams are lower than 38 °C. The fire protection system also needs to be more robust for waste streams having flashpoints up to 60 °C. In this business model case study, the facility design is considered for flashpoint more than 60 °C.

Table 13.19 Typical cost of a Liquid Co-processing facility

S. No	Item	Capacity	Cost in USD
1	Liquid co-processing system consisting of 40 M3 storage tank, filters, unloading pumps, feeding pumps, piping and nozzle	2 TPH	280,000
2	Laboratory and firefighting	As required	70,000
	Total		350,000

II.   Business model for co-processing of Liquid AFRs

Figure 13.7 depicts the business model for co-processing of liquid hazardous wastes from industries. These liquid waste streams get received from the industries directly or through a pre-processing agency.

III.   Salient features of the business model

Table 13.20 provides the salient features of the business model.

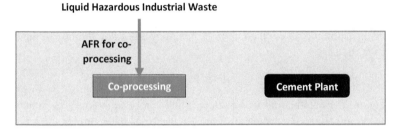

Fig. 13.7 Business model for Co-processing of industrial Liquid hazardous wastes

Table 13.20 Salient features of the business model of Co-processing of liquid AFR

S. No	Parameter	Value
1	Investment in co-processing facility (Table 13.19)	USD 350,000
2	Cost of coal (From Table 13.1)	USD 0.014 / 1000 kcal
3	Negotiated price of liquid AFR (Assumed)	-USD 14 / T
5	Co-processing cost (Typical)	USD 4.2 / T
6	Total cost	-USD 9.8 / T
8	NCV of liquid AFR	1000 kcal / Kg
4	Total cost of liquid AFR co-processing	-USD 0.0098 / 1000 kcal
7	Savings benefit of using Liquid waste (0.014 + 0.0098)	USD 0.0238 / 1000 kcal
9	Liquid waste used per day	40 TPD
10	Liquid Waste used per year (300 day / year)	12,000 Tons per year
11	Total energy from Liquid waste	12,000,000,000 kcal
12	Savings benefit derived by using Liquid waste	USD 285,600 per year
13	Payback Period of investment	1.23 years

13.5.6   Co-processing of Alternative Raw Materials Through Raw Meal

There are several waste materials that can be utilized to replace the natural raw materials such as lime sludge from paper industry, red mud from aluminium industry, mill scale from rolling mills, ETP sludge from water treatment plants, etc. As these waste streams do not contain any organic constituents, they can be co-processed in the kiln by feeding them along with the raw materials. They are waste material and hence it is cheaper also and therefore promoting appropriate business case.

These materials along with other raw materials get pulverized to the desired extent in the existing raw mill, become part of the raw meal, and get consumed in the cement manufacturing process. For undertaking the co-processing of these materials, the existing infrastructure is good enough. Hence, they add value directly to the bottom line of the industry.

I.   Example—Co-processing of mill scale

There are many industrial waste streams that can be utilized as raw materials in cement manufacture. Examples are lime sludge from paper mill, mill scale from rolling mills, red mud from Aluminium industry, ETP sludge from water treatment plant, steel slag from alloy steel industry, iron sludge from chemical industry, phosphate sludge from automobile industry, etc.

These materials can be directly co-processed in the kiln through the raw material or the calciner route. The raw material route is acceptable only when there are no volatile organics present in them. Volatile organics tend to get volatilized at the preheater top when they encounter high temperature of $> 350\ °C$ and cause VOC emissions which are not acceptable.

In the business model of co-processing such industrial wastes, the main driver is the natural raw material of cement manufacture that it will be replacing. The main natural raw materials that the cement industry considers replacing are lime, iron oxide, aluminium oxide, and silica, which are utilized in large quantities. As the feeding of these materials is generally done through the existing raw meal route, there is no specific capital investment required. There would be some handling cost but it would be nullified because of the replacement of the natural material.

Mill scale is a waste that is generated in the steel rolling mills. This has high iron content in the range of 80 to 95%. This can be easily utilized by cement plants as a replacement for iron ore or laterite which is normally used by the cement industry. Further, this material can be fed into the kiln along with the raw materials by proportioning it appropriately in the raw mix design.

II.   Business model for co-processing of raw materials

Figure 13.8 depicts the business model related to use AFR as an alternative raw material.

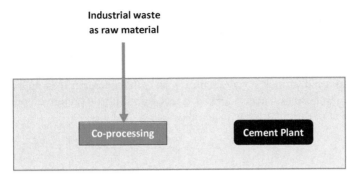

Fig. 13.8 Business model for Co-processing of industrial waste as Raw material

Table 13.21 Salient features of the business model of use of industrial waste as raw material

S. No	Parameter	Value
1	Investment in co-processing facility	USD 0
2	Price of laterite used in the cement plant	USD 8.4 / T
3	Iron content in laterite	40%
5	Cost of Iron from laterite (=8.4/0.4)	USD 21 / T
6	Negotiated price of mill scale	USD 11.2 / T
8	Iron content in mill scale	88%
4	Cost of Iron from mill scale (=11.2 / 0.88)	USD 12.73 / T
7	Savings benefit of using mill scale (21–12.73)	USD 8.27 / T Iron
9	Iron used in cement plant	200 TPD
10	Iron used per year (300 day / year)	60,000 Tons per year
11	Total savings using mill scale / year (=60,000*8.27)	USD 496,200 per year

III. Salient features of the business model

The salient features of this business model are illustrated in Table 13.21.

13.6 Conclusion

It can be concluded from this chapter that by considering the appropriate parameters associated with the use of AFRs, the different business models can be success-fully implemented. The cement plant co-processing operating in business mode from almost the 1980s and the same has survived for more than four decades with successful results in spite of various business challenges. It can be observed from the case studies and business models that industrial hazardous and non-hazardous waste that comes with gate fee provides better viability considerations. It is important, therefore, to include some amount of these materials in the AFR mix. Apart from the economic viability, co-processing also helps in reducing GHG emissions, conserving natural resources, and creating a vibrant circular economy. The government of India and various governments of many other countries have appreciated the co-processing business models and have recognized it as a preferred option for the management of different waste materials.

Reference

Guidelines on pre-processing and co-processing of wastes in cement production, GIZ, Lafarge-Holcim & University of applied sciences and arts (2020). Northwestern Switzerland

Chapter 14
Global Status of Co-processing

14.1 Introduction

Cement manufacturing is a highly resource intensive process and utilizes a large quantum of fossil materials in its manufacture. It emits a large quantum of CO_2 during its process due to calcination of limestone, which is a raw material, and also due to firing of fossil fuels. Cement industry, therefore, has a large carbon footprint accounting for about 7% of the CO_2 released globally. It has therefore planned for cutting CO_2 emissions by 24% below the current levels by 2050 (Cement technology roadmap) (IEA, 2018). To be sustainable and remain so, the cement industry is undertaking several initiatives globally to reduce its carbon footprint. Co-processing is one of the important pillars in achieving the desired reduction in the carbon footprint.

The requirement of cement has been always rising all over the world for past several years. Its growth is linked to the GDP of the specific country. There are several factors that influence the cement demand. Peak per capita cement consumption comes as countries transition from rural to urban societies, and as urban infrastructure is progressively built. As countries modernize and urbanize, their cement consumption has historically increased from around 100 kg/capita in rural populations, to a peak of around 1500 kg/capita/yr. The long-term average for developed countries that sporadically build new infrastructure but mostly just renovate, of around 500 kg/capita/yr.

Generally, the more is the cement demand in a particular country, the more is the feasibility of improving co-processing volumes in that specific country. Figure 14.1 provides an overview of the development of the per capita consumption of cement country-wise of many countries.

Global Cement and Concrete Association (GCCA) represents about 30% of the total cement production of the world. It is guiding the implementation of sustainability initiatives of cement and concrete manufacturing industries worldwide. It also monitors and documents the relevant information related to the sustainability initiatives of the cement and concrete sector.

© The Author(s), under exclusive license to Springer Nature Singapore Pte Ltd. 2022   313
S. K. Ghosh et al., *Sustainable Management of Wastes Through Co-processing*,
https://doi.org/10.1007/978-981-16-6073-3_14

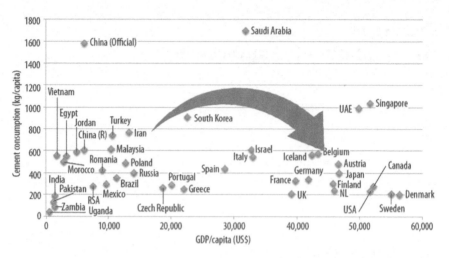

Fig. 14.1 Development of per capita cement demand (kg/capita) with increasing wealth (GDP in US$/capita) (Global Cement Magazine, May 2019).

It has created a database called "Getting Numbers Right" (GNR) containing vital data pertaining to the cement industry represented in GCCA. This includes data pertaining to cement production, CO_2 emissions, Power production and Consumption, Heat production and Consumption, and Mineral Components. This data does not represent the total cement industry of the world but represents a sizable portion of the same to provide a representative view of the sustainable growth pursued by the cement industry. This database is up to date till 2018. This data also contains the status on co-processing undertaken by the member companies which is utilized for undertaking the necessary assessment in this chapter.

14.2   Status of Co-processing in Cement Industry in Different Countries

To evaluate the global status of co-processing, the data available in the GNR database of GCCA is utilized (GCCA, 2019). This assessment is carried out with the objective of understanding the key success factors in implementing co-processing.

14.2.1   Country Average TSR% of Different Countries

The country average TSR% of cement industries of different countries are tabulated in Fig. 14.2. It can be observed from Fig. 14.2 that the data represents countries spread across the continents. These include several counties from Europe, Brazil from South

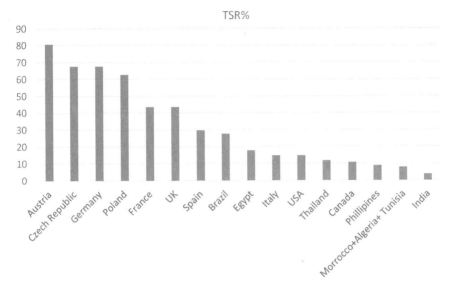

Fig. 14.2 Country average TSR% of cement industries

America, Thailand, Philippines, India from Asia, Egypt, Morocco, Tunisia, Algeria from the Middle East and North Africa, and Canada and USA. It should be noted that co-processing is implemented by cement companies from other geographical locations also such as China, several countries in the Middle east, Australia, New Zealand, South American countries, Japan, Asia and Southeast Asia, etc.

The data in Fig. 14.2 demonstrates that several countries in Europe lead the AFR drive in the world. Austria has the highest TSR% and India has the lowest TSR% among the countries representing the GCCA member companies. This graph also indicates that Austria has far less capacity available in the kilns to manage additional waste and India has substantially more capacity available to manage additional waste.

14.2.2 Volume of Alternative Fuels Co-processed by Different Countries

To review the global status of co-processing, another angle is considered wherein the volume of Alternative Fuels co-processed by countries is evaluated. Based on this assessment, the volume of AF utilized by cement plants in different countries in 2018 is tabulated in Fig. 14.3. It can be observed from Fig. 14.3 that Germany stands out with the highest volume of AF utilization and Canada has consumed the lowest volume. India occupies 5th position in terms of AF utilization and Austria occupies 11th position.

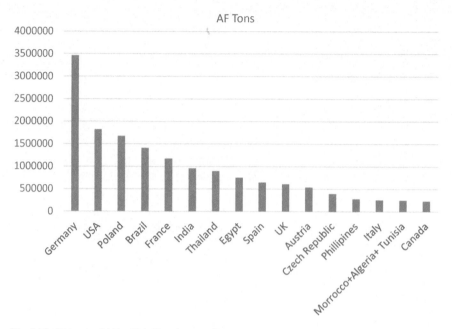

Fig. 14.3 Volume of AF utilized by cement plants

14.2.3   *Additional Evaluations on the Global Status of Co-processing*

Following additional evaluations are carried out from the data available in the GNR:

A.   Biomass utilization in terms of TSR%,
B.   Biomass utilization in terms of Volume of AF co-processed,
C.   Utilization of other AF in terms of TSR%,
D.   Utilization of other AF in terms of Volume co-processed.

The analysis of the results from this assessment indicates to us the following:

a.   The Czech Republic leads the case of highest percent substitution of fossil fuels with Biomass and India trails with the lowest percent utilization.
b.   In terms of volume utilization of Biomass, Germany tops the list and Italy occupies the lowest position.
c.   Utilization of other AFs in terms of TSR% is led by Austria and the lowest utilization rate is that of Philippines.
d.   Co-processing of the maximum volume of other AF is carried out by Germany and the lowest volume of AF is co-processed by Philippines.

The analysis results of the evaluations as above are tabulated in Figs. 14.4, 14.5, 14.6, and 14.7, respectively.

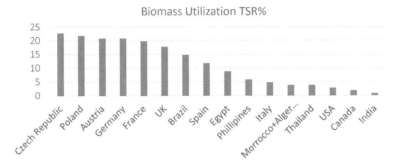

Fig. 14.4   Biomass utilization TSR%

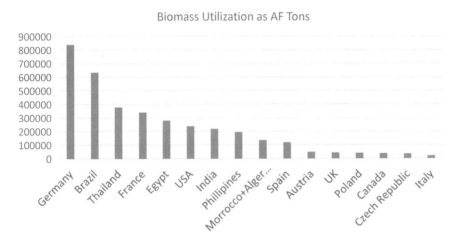

Fig. 14.5   Biomass utilization AF tonnes

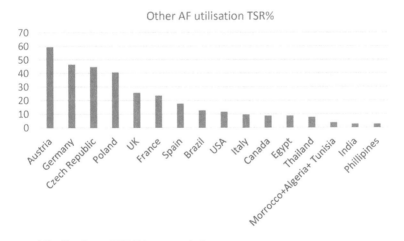

Fig. 14.6   AF utilization as TSR% in cement industry

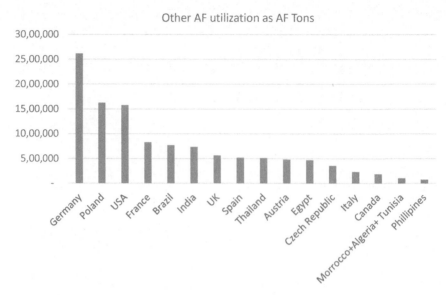

Fig. 14.7 Biomass utilization AF tonnes

14.3   Country Specific Status on Co-processing

Having reviewed the co-processing data of the cement plants in different countries worldwide, it would be interesting to review the same figures of the cement industry in specific countries to review how co-processing has progressed in these countries over a period of time. For illustration purpose, Austria, USA, Brazil, India, Egypt, and Poland located in different parts of the world are considered for evaluation whose average TSR% data of the cement plants are demonstrated through Figs. 14.8,14.9,14.10, 14.11, 14.12 and 14.13, respectively.

Following observations are made from the above figures:

a.   The TSR of the cement plants in all countries have grown on a yearly basis.
b.   The rate of growth of TSR in some countries is significantly higher than the others.
c.   The rate of growth of TSR% varies from country to country. It depends upon the legislation, availability of the waste materials in the market place, and drive towards sustainability.
d.   The TSR% also depends upon the economics of the utilization of AFRs in the cement kilns.

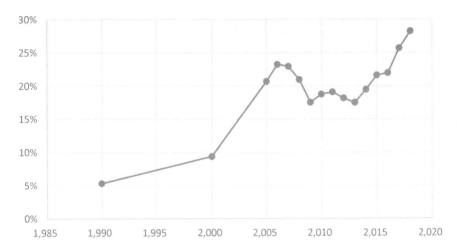

Fig. 14.8 TSR% growth trend-Austria

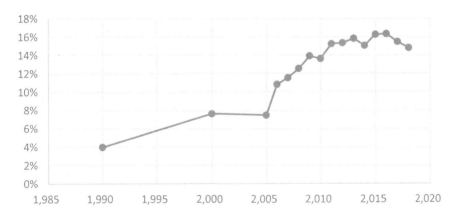

Fig. 14.9 TSR% growth trend-USA

14.4 Conclusion

It is understood from the deliberations in this chapter that co-processing is practiced successfully by the cement industry in many countries globally. Different countries have substantially different growth rates in achieving TSR%. The drive for implementing co-processing depends upon the local conditions. There are quite a few countries that have achieved an average TSR% figure of more than 70%. This indicates that the countries, that are currently operating at low TSR%, have substantial capability available in them to play a major role in facilitating sustainable management of wastes.

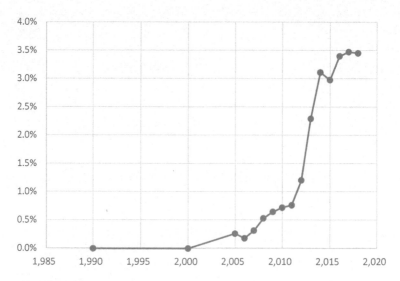

Fig. 14.10   TSR% growth trend-India

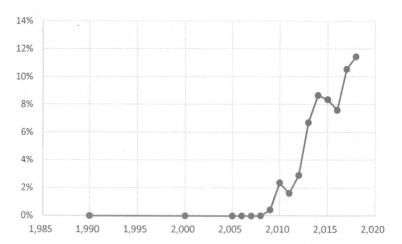

Fig. 14.11   TSR% growth trend-Egypt

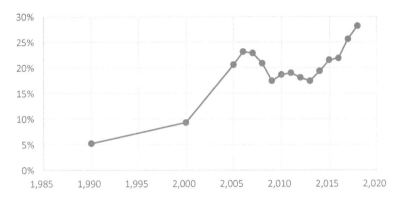

Fig. 14.12 TSR% growth trend-Brazil

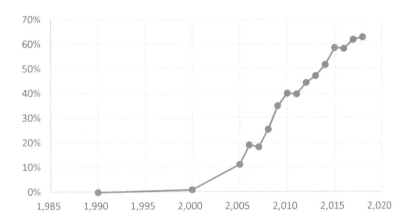

Fig. 14.13 TSR% growth trend-Poland

References

IEA. Cement technology roadmap plots path to cutting CO_2 emissions 24% by 2050 - News (2018)
GCCA, GNR data base 2019. available at https://gccassociation.org/gnr/. Global Cement Magazine, May 2019

Chapter 15
Journey of the Growth of Co-processing in India

15.1 Introduction

Considering increasing waste production on the one hand and the resource demands of the cement process on the other hand, cement companies started in 1979 to look at waste as a source of raw material and energy (WBCSD, 2014). The Indian cement industry was aware of the success achieved by the cement industries in some of the developed countries in utilizing wastes as alternative fuels. However, it was not able to initiate tangible actions until 2005 because of the lack of sufficient technical knowledge and because of the lack of facilitating regulatory framework to implement the same. A few of the plants during that time were utilizing biomass as AFRs.

Currently, many cement plants in India are implementing co-processing as a sustainability initiative in a business mode. An assessment carried out by GCCA under cement sustainability initiative indicated that the amount of waste material co-processed as Alternative Fuel in 2018 by its member cement companies based in India was more than 0.7 million TPA. This had resulted in a Thermal Energy Substitution rate of ~ 4%. The quantum of waste material co-processed as alternative raw material by the same companies in India was not assessed by GCCA but would be certainly adding to the volume co-processed in the country. While traversing this journey, the cement industry had to overcome several challenges and go through a learning curve.

These learnings and challenges were in the following areas:

1. Building awareness of the stakeholders on the concept of co-processing,
2. Implementation of appropriate regulatory provisions favouring co-processing,
3. Capability development of the cement industry on the technological aspects of co-processing,
4. Capacity development of the authorities on the regulatory provisions required for facilitating sustainable co-processing,
5. Development of the waste market for implementing co-processing in a sustained manner, and
6. Achieving desired regulatory framework that promotes co-processing.

© The Author(s), under exclusive license to Springer Nature Singapore Pte Ltd. 2022
S. K. Ghosh et al., *Sustainable Management of Wastes Through Co-processing*,
https://doi.org/10.1007/978-981-16-6073-3_15

To achieve the desired success in the above-defined areas, stakeholder engagement and policy advocacy are two important tools. Considerable efforts were put in by several agencies in India in implementing these tools. The concepts of stakeholder engagement and policy advocacy are properly defined in the next sections, and then the journey traversed by the Indian cement industry in respect of achieving the desired objective is documented in this chapter.

15.2   Stakeholder Engagement and Policy Advocacy

Effective stakeholder engagements are vital to achieving smooth and trouble-free operations in AFR co-processing. Many stakeholders are important such as cement plant executives, authorities, the community, waste generators, and waste management agencies.

Appropriate advocacy is essential to ensure that the right practices that facilitate positive impacts on the environment and stakeholders prevail and sustain. The important stakeholders that need to be addressed in the advocacy initiatives are governmental authorities, NGOs, activists, industry associations, environmental testing agencies, etc.

15.2.1   Stakeholder Engagement

Stakeholder engagement involves designing a proper communication plan based on the critical assessment of the current situation. This engagement plan involves continual interaction with the stakeholders with appropriately designed messages and communications. These messages and communications play an important role in clarifying the perceptions and mindsets of the stakeholders so that the concerns are clarified and mitigated. These interactions are required to be carried out through well-designed engagement processes.

The design of an appropriate engagement plan consists of the following activities.

A.   **Critical Assessment of the current situation**
     Most of the cement plants will be having community engagement activities under corporate social responsibility (CSR). Corporate social responsibility (CSR) refers to strategies that companies put into action as part of corporate governance that is designed to ensure that the company's operations are ethical and beneficial for society. The initiatives under CSRA generally address the issues of the society related to the environment, human rights, philanthropic and economical aspects.
     While implementing CSR initiatives, it is important to learn about the concerns that stakeholders have in respect of co-processing. This learning can be carried

out in-house or through the engagement of a third-party agency that is qualified in implementing such assessments. This assessment should help understand the mindsets and perceptions of the stakeholders, map them properly into appropriate categories, and project their impacts on various fronts.

B. **Designing engagement plan**

Stakeholder engagement needs to be designed while ensuring that trust prevails with the stakeholders. For trust to prevail, it is desired that complete transparency is maintained with the stakeholders by providing them the required data and information with complete openness. The process needs to be designed for continual engagement by sharing with them the long-term objectives and future plans. A clear communication plan also needs to be put in place to bring clarity into the roles and responsibilities of the executives involved in implementing stakeholder engagements.

C. **Implementing engagement plans**

The engagement plans can be implemented through various options such as meetings, regular mail communications, FAQs, standard presentations, consistent messages, brochures, case study illustrations, conferences, community advisory panels, and site visits, etc.

All messages and communications need to be put on a continuous improvement process. For this, an appropriate communication improvement process needs to be put in place so that the messages go through the desired refinement as shown in Fig. 15.1.

The following are the important tips on implementing stakeholder engagement and dialogues effectively.

1. Understand the subject matter thoroughly so that the same can be communicated to the stakeholder and the desired objective can be accomplished.
2. Stakeholder views are important and need to be listened to attentively.
3. Do not bring complexity in the communication. It is always better to communicate in a simple language.

Fig. 15.1 Communication improvement process

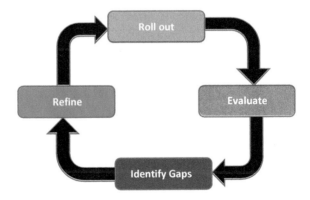

4.   Irrespective of the results of the interaction, always ensure that at the end the door is still open for another interaction later on when needed.
5.   It is always good to express your gratitude to the stakeholder.

An objective-driven approach is desired in implementing the stakeholder engagement so that the engagement process becomes effective. Further, it would be good to build tangible partnerships with the stakeholders to improve effectiveness in the engagements.

The following are the typical examples of stakeholder engagements for co-processing.

a.   Formation of the cooperative federation of the farming community for the management of their non-cattle feed crop residue.
b.   Collaboration with local administration for achieving a clean and healthy living place.
c.   Plastic waste segregation and management through community collaboration.
d.   Formation of the Community Advisory Panels to address the stakeholder concerns.

15.2.2   Policy Advocacy

Co-processing deals with wastes that are controlled through legal processes. As wastes need to be managed in a sustainable manner and co-processing facilitates the same, it is desired that the legal processes are fully aligned with the objectives of co-processing. To achieve this objective, robust advocacy efforts are desired to be implemented. Successful advocacy requires meticulous planning, desired stakeholder engagement, and commitments towards collaborative effort. In the policy advocacy initiatives, the following activities are involved.

A.   **Assessment of the existing policy framework**
     Advocacy is carried out to implement desired changes in the regular framework of the country. It is therefore desired that the existing policy framework is reviewed properly to understand the implications of the proposed changes on environmental issues, ease of operation, health and safety, legal and social considerations, etc. Towards this, preparing position papers on the relevant topic helps a lot. This position paper becomes very handy to the authorities when they take up the cause of working on the concerned areas. It is also very important to understand the perceptions and mindsets that the stakeholders are carrying with them so that they can be addressed through different methodologies such as documented FAQs, query-clarification dialogues, illustrative messages, practical demonstrations, and illustrations on the benefits of the proposed changes.

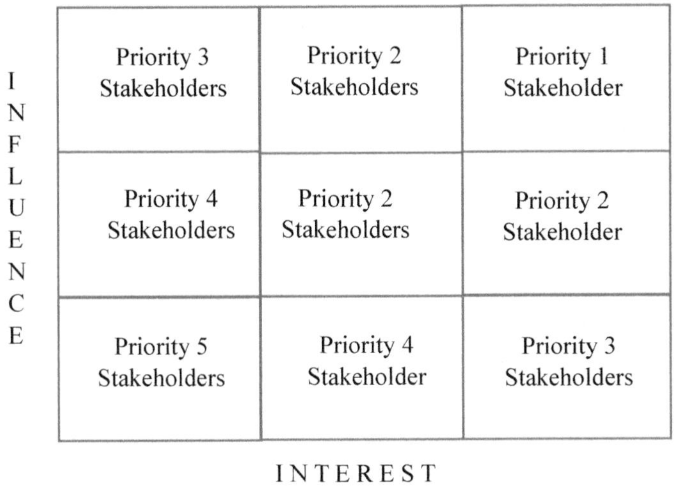

I	Priority 3 Stakeholders	Priority 2 Stakeholders	Priority 1 Stakeholder
N **F** **L** **U** **E**	Priority 4 Stakeholders	Priority 2 Stakeholders	Priority 2 Stakeholder
N **C** **E**	Priority 5 Stakeholders	Priority 4 Stakeholder	Priority 3 Stakeholders

I N T E R E S T

Fig. 15.2 Mapping of stakeholders

B. **Mapping of Stakeholders**

Implementation of advocacy requires a well-documented advocacy plan, effective stakeholder engagement, impactful messages, and appropriate communication strategy with effective messages. For achieving desired success in advocacy, it is desired that all the stakeholders are properly identified and mapped in a two-by-two matrix with their interest in the advocacy objective as one axis and their influencing capability as another axis as shown in Fig. 15.2.

As shown in the diagram, the stakeholders are required to be motivated to become priority 1 stakeholders and shift to the top-right corner so that they become ambassadors of the advocacy implementation process. This transformation to priority 1 stakeholders is possible through the stakeholder dialogue and engagement processes. The engagement can be through a one-to-one dialogue and/or through a multi-stakeholder deliberation process.

C. **Communication with the Stakeholders**

When a one-to-one approach is carried out, it is desired to communicate with the stakeholder the shortcoming in the existing policy framework and share alternate solutions to deal with them. It is also desired to share with him appropriate messages that are designed in the form of brochures or leaflets or hand-outs so that the message remains with him for a long period. One person obviously cannot bring about the required modification in the policy framework, but he becomes an important ambassador to voice out the suggested modifications to the relevant stakeholders.

When multistakeholder deliberations are desired, it is better to carry them out through an industry association. Several methodologies can be implemented such as the following:

1. Stakeholder dialogues
2. Round table conference
3. Stakeholder conference
4. Forum of dedicated stakeholders
5. Stakeholder missions for learning and demonstrations
6. Expert committees
7. Social media campaigns
8. Communications through Print and Digital media
9. Personalized letters specifically addressed to the relevant authority,
10. Etc.

The various issues that need to be addressed through advocacy are the following:

a. Utilization of maximum quantum of AFRs
b. Free movement of AFRs across borders
c. Polluter pays principle
d. Respect of waste management hierarchy
e. Avoidance of dilution
f. Emission standards

Several efforts were implemented by various agencies towards stakeholder engagement and advocacy to boost the co-processing initiative. With these efforts, desired objectives were achieved by the cement industry. To elaborate this journey appropriately, it is divided into two different time periods—2003–2008 and 2008–2020.

15.3   Journey—2003–2008

Co-processing is defined as a technology of utilizing wastes as resources in different resource intensive industries (RII). Although wastes can be utilized in some resource-intensive industries such as cement, power, steel, lime, glass, and refractories, the use of wastes in cement manufacture dominates the RII sector. The specific physico-chemical features of the cement manufacturing process allow this to happen at very high scales. This option of managing wastes from Municipal, Industrial, and Agriculture sectors as alternative fuels and raw materials in the manufacture of cement has been in practice globally since the 1980s, and many cement plants worldwide are exercising this option at different scales of operation.

A need was felt in India in 2003 to promote the utilization of hazardous combustible waste having a higher calorific value in cement kiln as fuel (CPCB, 2004). This need was to promote utilization of hazardous combustible waste having a higher calorific value in cement kiln as fuel. In the earlier times, the operation of using wastes as alternative fuels in cement kiln used to be called co-incineration. Later, it was corrected to co-processing. It was considered that this will not only solve the disposal problem associated with hazardous waste but also conserve natural fuel resources.

There were, however, stakeholder concerns about the possibility of emissions of toxic metals, volatile organic carbon compounds, and other toxic gases, while undertaking co-processing of wastes in cement kilns. There was also concern about the possibility of a negative impact on the quality of cement. Therefore, there was a need to evaluate the sanctity of this stakeholder concern by conducting the trial. In such a trial, the measurements of the identified emission parameters are carried out in three stages.

1. Before initiating co-processing of AFR (Pre-trial),
2. During co-processing of AFR (During Trial), and
3. After co-processing of AFR (Post Trial)

During this trial, various identified emission parameters are measured by sampling the kiln exhaust gas as per the defined procedure. The testing of quality parameters of the clinker and cement produced as per the desired quality considerations is a part of each stage. CPCB in India has prepared a step-wise approach to evaluate the same through trials to be carried out in cement plants as per a well-designed trial procedure with an emission monitoring protocol (CPCB, 2004). This CPCB's emission monitoring protocol for the demonstration trial is provided in Table 15.1.

Table 15.1 Emission monitoring schedule during a trial run of co-incineration of hazardous waste in cement kiln

S. No.	Parameter	Frequency	Monitoring agency
1	Particulates	4 samples/day	External Lab/CPCB
2	SO2	4 samples/day	External Lab/CPCB
3	HCL	4 samples/day	External Lab/CPCB
4	CO	4 samples/day	External Lab/CPCB
5	NOx	4 samples/day	External Lab/CPCB
6	TOC	1 sample/day	External Lab/CPCB
7	HF	4 samples/day	External Lab
8	Hydrocarbons	2 samples/day (of four hours each)	External Lab
9	Opacity	Continuous	Cement plant
10	VOC	2 samples/day	External Lab
11	PAH	2 samples/day	External Lab
12	Heavy Metals (Cd, Th, Hg, Sb, As, Pb, Cr, Co, Cu, Mn, V, Zn, Sn, Se)	1 sample/day	External Lab
13	Dioxin and furans	1 sample/day	GTZ, Delhi

Table 15.2 Co-incineration trials (21-day duration) conducted by different cement companies in India

S. No.	Waste material	Name of cement plant
1	Chemical industry ETP Sludge	M/s Rajashree Cement, Gulbarga, Karnataka, a unit of Grasim Industries Group (Now Ultratech Cement Ltd.)
2	Refinery sludge	M/s Grasim Cement, Reddipalayam, Tamil Nadu (Now Ultratech Cement Ltd.)
3	Paint sludge	-do-
4	Used tyre chips	-do-
5	CETP sludge	M/s J K Laxmi Cement, Sirohi, Rajasthan
6	TDI Tar	M/s Ambuja Cement Ltd., Kodinar, Gujarat

This monitoring was suggested during the trial that consisted of three phases of one week each: (1) Pre-trial, (2) During Trial, and (3) Post trial.

With this trial protocol, the following trials were carried out by CPCB and the cement plants as listed in Table 15.2 after receiving necessary permissions from the respective State Pollution Control Boards.

The results of these trials had been very encouraging in concluding that co-processing of these different waste streams did not influence the emissions from the cement kiln or the quality of the cement product in any significant manner (CPCB, 2006). During the period 2003–2006, this process of utilization of wastes as alternative fuels in cement kilns was referred to as co-incineration. However, in the subsequent deliberations, it was appropriately referred to as co-processing because of the following considerations:

1.   Cement kilns do not incinerate waste materials but utilize them as input resources.
2.   Waste materials not only get utilized as alternative fuels but also get utilized as alternative raw materials. They are therefore referred to as alternative fuels and raw materials (AFRs).

All these trials demonstrated that there was no major impact on the emissions from cement plants. Subsequently, it was decided that the permission for co-processing as a waste stream would be granted to a cement plant only after it carries out the 21-day demonstration trial of that waste stream as per the defined protocol. The cost of implementing such a trial was observed as prohibitively expensive and hence there was hesitation in the cement industry to undertake the co-processing initiative.

In 2005, Holcim (Switzerland) took over "The Associated Cement Ltd. (ACC)" and "Ambuja Cement Ltd. (Ambuja)". Holcim, having co-processing of AFRs as one of the important business pillars, introduced the co-processing concept and provided the technological inputs to ACC and Ambuja to initiate it in the respective organizations. Subsequently, ACC and Ambuja formally initiated the co-processing activities in 2006 in India under the brand name "geocycle".

The major effort required in India at that time was to bring appropriate awareness of the stakeholders on the concept of co-processing and promote it in a manner that will facilitate the smooth implementation of co-processing in the Indian cement industries.

From 2006 to 2007, ACC, Ambuja, and Holcim carried out several stakeholder meets in different states in India in association with GTZ (then GIZ), CPCB, and respective SPCBs. The objective of these meets was (a) to bring appropriate awareness of the stakeholders on the concept of co-processing; (b) to deliberate the advantages, challenges, and concerns related to co-processing, and (c) prepare appropriate recommendations for formalizing co-processing in the country as per the best practices being implemented globally. Figure 15.3 provides a photo of such a stakeholder meet organized in Chandigarh, Northern India in 2008.

Considering the relevance of including the provision of co-processing technology in the proposed revision in the Hazardous Waste Management Rules 2008 (HWM Rules 2008), the recommendations from all these stakeholder meets were submitted to the then authorities of Ministry of Environment and Forests (MoEF), Government of India, CPCB, and respective SPCB. Although the provisions of co-processing technology did not get included in the revised HWM Rules 2008, the new Rule 11 got included in the HWM Rules 2008. This rule mentioned that "The utilization of hazardous wastes as a supplementary resource or for energy recovery, or after processing shall be carried out by the units only after obtaining approval from the

Fig. 15.3 Stakeholder meet organized in Chandigarh 2008

Central Pollution Control Board" (HWM Rules 2008). This rule opened the opportunity for implementing co-processing of wastes as alternative fuels and raw materials in cement manufacture.

15.4   Journey 2008–2020

Considerable efforts were put in the country to promote and implement co-processing for achieving sustainable management of wastes from Industrial, Municipal, and Agricultural sectors. The impetus to utilize a larger quantum of AFR in the Indian cement industry started with the identification of AFR as one of the important levers for CO2 reduction in the low carbon technology road map prepared by the Indian cement industry.

   These efforts include the following considerations:

(a)   Technology Demonstration for environmentally friendly management of wastes.
(b)   Stakeholder Awareness and advocacy on the features of co-processing technology for including co-processing in the Indian policy framework.
(c)   Notification of the Rules, Guidelines, Emission standards, and monitoring protocol.
(d)   Status on co-processing in India.

15.4.1   Demonstration of Co-processing Technology for the Sustainable Management of Wastes

Rule 11 of HWM Rules 2008 required cement plants to demonstrate that the cement kiln co-processing technology is environmentally safe for managing hazardous and non-hazardous wastes. The major hurdle was the high cost associated with implementing the trial as per the 21-day monitoring protocol. CPCB reviewed this 21-day protocol by evaluating the results of the trials conducted on different waste materials in different cement plants and then modified the 21-day monitoring protocol to 5-day monitoring protocol. This new monitoring included 1-day pre-trial, 3 days during co-processing, and 1-day post trial.

   This new protocol was reasonably affordable to the cement plants and also to the industries generating wastes. Hence, about 88 co-processing trials were conducted by different cement companies from 2008 to 2016. These trials were carried out on different kinds of hazardous and non-hazardous wastes.

   Table 15.3 lists out the number of trials implemented by different companies in their plants spread across the country.

   CPCB analysed the results of the monitored data of these trials and confirmed that co-processing is not influencing the emissions from the cement kiln in any adverse manner while co-processing the wastes. CPCB then approved these waste streams

Table 15.3 Co-processing trials (5-day duration) conducted by different cement plants in India

S. No.	Cement industry	Number of trials
1	ACC Limited	33
2	Ambuja Cement Ltd.	20
3	Ultratech Ltd.	20
4	Lafarge India Pvt. Ltd.	7
5	Shree Cement Ltd.	4
6	Sanghi Cement Ltd.	1
7	My Home Cement Ltd.	1
8	J K Lakshmi Cement Ltd.	1
9	Zuari Cement Ltd.	1

and allowed their co-processing in cement plants after obtaining necessary permit from relevant SPCBs (CPCB, 2016).

One of the major achievements in promoting co-processing in the country was the successful co-processing trial of plastic waste that was conducted in March 2008 in ACC Kymore Cement Works. This was carried out in association with Indian Centre for Plastics in Environment (ICPE) and under the supervision of CPCB and MPPCB. This trial was carried out with a mix of different plastic waste materials (ICPE, 2008). Because plastic is a problem material and co-processing offered a positive and sustainable solution for its sustainable management, both CPCB and SPCBs started extending positive support to the cause of co-processing and a large number of cement plants started co-processing plastic waste by receiving it from ULBs.

The operating process for the trial-based permitting process is depicted in Fig. 15.4.

The timeline from obtaining permit based on this process used to be 1–2 years. Implementing more trials used to provide better opportunity to receive the permit for the waste stream first. The waste materials utilized in these trials consisted of hazardous and non-hazardous wastes from industrial and municipal sectors. Evaluations of the results of these trials have demonstrated that the wastes get managed in an ecologically sustaining and environmentally friendly manner.

Generally, wastes tend to have large variation in their chemical constituents, and there used to be many stakeholder concerns on the capability of the cement kiln to deal with such variation.

To address this concern, the constituent levels present in the waste streams approved for co-processing were evaluated. For this, the chemical constituents present in the approved waste streams were tabulated from the data available from 22 trials, and their minimum and maximum values were compiled. These values are tabulated in Table 15.4.

The results of this analysis concluded that the cement kilns are capable of dealing successfully with such large variation in the chemical constituents present in wastes. It also concluded that rather than the concentration, it is the absolute quantum getting

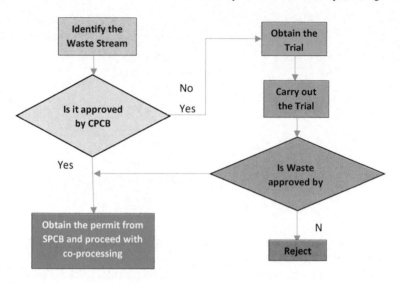

Fig. 15.4   Operating framework of the trial-based permitting system

into the kiln that is more important to be monitored (Ulhas Parlikar, 2016). Apart from the co-processing trials implemented in India to demonstrate the successful management of different hazardous and non-hazardous waste materials, a trial for the demonstration of the disposal of complex material such as ozone depleting substance (ODS) was also implemented.

Persistent organic pollutants (POPs) and the ozone depleting substances (ODS) are two highly complex and difficult to dispose materials. While demonstrating the disposal of such materials, ensuring satisfactory destruction and removal efficiency (DRE) is must. This trial was conducted with the involvement of MoEF, CPCB, MPPCB, GPCB, academicians, etc., in the Kymore Cement Works of ACC Ltd. For emission monitoring, the international agencies were involved.

The trial was successfully implemented with the DRE of > 99.9999% and the overall environmental performance in compliance with Indian regulation and international best practice (Karstensen, 2014).

Figure 15.5 depicts the banner on the facility set up for co-processing CFCs in Kymore plant.

The results of many of the trials implemented in India have been analysed in detail by several agencies in detail, and the assessments have been published in the various national and international scientific journals. A detailed list of these publications has been provided in the references section.

AFR usage in the Indian cement industry in 2008 was almost negligible which improved to about 4% by 2016. Less than 10 cement plants were co-processing AFRs in 2008 which increased to about 60 plants by 2016.

Table 15.4 Range of chemical constituent variation present in the waste materials approved for co-processing

Parameter	Units	Min	Max
Moisture	%	0.60	67.40
Ash	%	0.96	98.70
VM	%	0.30	94.90
FC	%	0.10	45.70
Carbon	%	0.40	75.60
Hydrogen	%	0.20	9.10
Nitrogen	%	0.00	15.50
Sulphur	%	0.10	22.00
Oxygen	%	0.00	76.30
GCV	Kcal/Kg	80.00	7960.00
NCV	Kcal/Kg	114.80	6042.00
Mineral matter	%	3.50	34.50
Chloride as Cl	mg/Kg	0.00	14,200.00
Fluoride as F	mg/Kg	0.00	20.10
Cadmium	mg/Kg	0.10	27.60
Chromium	mg/Kg	0.20	36,229.70
Copper	mg/Kg	1.00	8,848.60
Cobalt	mg/Kg	0.10	176.40
Manganese	mg/Kg	0.10	7,800.00
Nickel	mg/Kg	0.10	9,300.00
Lead	mg/Kg	0.10	633.30
Zinc	mg/Kg	1.00	22,000.00
Arsenic	mg/Kg	0.10	8.10
Mercury	mg/Kg	0.10	3.80
Selenium	mg/Kg	0.00	8.20
Antimony	mg/Kg	0.10	9.40
Vanadium	mg/Kg	1.00	82,400.00
Thallium	mg/Kg	0.10	1.00
Tin	mg/Kg	0.00	145.60
VOC	mg/Kg	4.20	207.00
SVOC	mg/Kg	BDL	0.20
PCB	mg/Kg	0.00	0.50
PCP	Mg/Kg	BDL	1.40
TOC	%	0.00	66.00

Fig. 15.5 Banner on the Facility set up for co-processing of CFCs in ACC Kymore Cement Works along with authors Dr. Kare Helge Karstensen and Mr. Ulhas V. Parlikar

15.4.2   Stakeholder Awareness and Advocacy Initiatives on Co-processing Technology

To bring awareness of the stakeholders on the concept and capability of the cement kiln co-processing technology and to achieve its inclusion in the waste-related policy framework of India, various initiatives were taken up by many different agencies. Initiatives in respect of this awareness generation and advocacy of a few of the prominent agencies are mentioned below.

A.   Confederation of Indian Industries (CII)

CII is a non-governmental, not-for-profit, industry-led organization that plays a proactive role in India's development process. CII-Sohrabji Godrej Green Business Centre (CII-Godrej GBC) is a division of CII and offers advisory services in different areas. CII in association with Cement Manufacturers Association (CMA) had initiated a project to facilitate use of wastes as alternative fuels and raw materials in the Indian cement industry. This project was supported by M/s Shakti Sustainable Energy Foundation (SSEF), a part of Climate Works Foundation.

   Stake holder awareness

   i.   AFR deliberations in the Green Cementech conference.

ii.   Round table conferences in association with State Pollution Control Boards.
iii.   AFR Missions of the stakeholders representing cement plant executives and other stakeholders to various operating installations of co-processing in the cement plants in India.
iv.   Visit of authorities from CPCB and SPCBs to pre-processing and co-processing installations in India, Asia, and Europe to make them appreciate the different scales at which co-processing is implemented in these regions and the governing policy framework for the same.
v.   Parallel AFR conference in the Waste Management Summit which is an annual event organized by CII-Godrej GBC for deliberations on the AFR-related subjects.
vi.   AFR-relevant best practices sharing sessions in different events organized by CII-Godrej GBC.
vii   Awareness sessions at different locations in the country through the regional centres of CII.
viii. Awareness generation sessions by visiting cement plants located in different clusters.
ix.   Cluster-wise awareness sessions on AFR for the benefit of ULBs and industrial organizations.
x.   B2B meetings between generators, waste processing industries, and cement plants.
xi.   Implementing waste exchange website for facilitating the business tie-ups between waste generator and cement plants.
xii.   Publishing literature on the relevant aspects related to AFR.

- Case study manual on alternative fuels and raw materials utilization in Indian cement industry (2011).
- Promoting alternative fuel and raw material usage in Indian cement industry.
- Co-processing: Journey Towards Low Carbon Economy—Pillar to support India's Intended Nationally Determined Contribution (INDC).
- Reports of the missions—national and international.
  The various initiatives taken up by CII-Godrej GBC are depicted in Figs. 15.6 and 15.7.

Advocacy

- Constitution of an expert group representing officers from CPCB and SPCBs of eight states. These eight states were Tamil Nadu, Maharashtra, Karnataka, Andhra Pradesh, Gujarat, Madhya Pradesh, Chhattisgarh, and Rajasthan.
- Demonstration of the environmentally sound features of the co-processing technology to the expert group by visiting plants in India and Asia and Europe.
- Exposure of the expert group to the policy frameworks of different countries related to co-processing.
- Interaction of the expert group with the authorities in different countries on the policy framework related to co-processing.

Fig. 15.6 Awareness and Advocacy initiatives by CII-Godrej GBC. Top: Mr. Ulhas V. Parlikar, author of the book addressing the seminar

- Submission of the "While Paper on Increasing co-processing in Indian Cement Plants" in 2012 (CII, 2012) prepared by the expert group.
- Preparation of a draft of an amendment required for incorporating co-processing in the policy framework and Chhattisgarh submission of the same to MoEFCC and CPCB.
- Organization of the stakeholder meets to deliberate the drafts of the policy framework prepared by MoEFCC and submitting the recommendations to MoEFCC.

B.   Cement Manufacturers Association (CMA)

Cement Manufacturers Association (CMA) is the apex body of large cement manufacturers in India. Representing almost ~ 90% of the total installed cement capacity in

Fig. 15.7 Awareness and advocacy initiatives of CII-Godrej GBC

India, CMA endeavours to create an enabling business environment for the Cement Industry in India. CMA engages in policy and regulatory level discussions with the Central/State Governments on matters and concerns raised by its Member Companies. CMA is part of various committees/working groups constituted by the Government of India to find sustainable solutions towards efficient utilization of waste by the Cement Industry. CMA organizes its flagship event "CONSERVE" biennially, inviting policy makers, academia, sector experts, and Industry representatives to deliberate over enhanced and effective utilization of waste in co-processing. CMA partnered with the Government of India in 2019 for one of India's flagship campaigns Swachhata hi Seva on minimizing single-use plastic within the country. During this limited duration campaign, CMA Member Companies reached out to 17 states, 377 villages, covering more than 1,00,000 people through 500 + activities. The project also involved the disposal of almost 8000 tons of plastic waste over about 3 weeks by the CMA Member Companies at their plants in close coordination with urban local bodies. CMA was awarded by the Ministry of Jal Shakti, Government of India, for its contribution to the initiative.

C. Geocycle India

Geocycle India is a shared services department of ACC Ltd. and Ambuja Cement Ltd. Geocycle India is a part of the global Geocycle network and has been in operation in the country for more than a decade. It is involved in pre-processing of wastes into AFRs and then co-processing them into cement kilns.

Stakeholder awareness

- Organization of stakeholder meets with SPCBs.

- Making presentations in the events organized by local/regional industry associations, national conferences, etc.
- Publishing articles in national/international journals and publications.
- Media interactions for promoting awareness on AFRs.

Advocacy

- Advocacy initiatives individually and in association with various industry associations.
- Advocated for achieving preferred status to the technology of co-processing in the regulatory framework and also permitting interstate movement of hazardous wastes.
- Participated as an expert in the committees constituted by CPCB and MoEFCC and prepared draft guidelines for the pre-processing and co-processing of waste streams.
- Showcased the business model of the co-processing initiative to the Task Force Committee constituted by CPCB to demonstrate the sustainability of the co-processing initiative in the country.

D.   GIZ (Earlier GTZ)

Deutsche Gesellschaft für Internationale Zusammenarbeit (GIZ) GmbH is working jointly with partners in India for sustainable economic, ecological, and social development. The thematic areas of GIZ in India are Energy; Environment, Climate Change and Biodiversity; Sustainable Urban and Industrial Development, and Sustainable Economic Development. GIZ partnered with Holcim in 2006 in bringing out the guidelines on co-processing in 2006 (GIZ/Holcim, 2006)and later with LafargeHolcim bringing out a refined version of the same in 2018 (GIZ/LafargeHolcim, 2018).

Stakeholder awareness

- Organizing interactive meets and deliberations on aspects related to AFR.
- Promoting co-processing through stakeholder meets as well as town-specific solid waste management improvement projects.
- Promoting co-processing through demonstration projects: Conducting trial run for dry fractions of municipal Solid Waste with Nashik Municipal Corporation and Geocycle as a pilot.

Advocacy

- Prepared a report on assessment and mapping of cement plants and municipal solid waste (MSW) processing facilities within 100, 200, and 300 km radius for promoting RDF co-processing in India (GIZ, 2018).
- Capacity building of the authorities in MoEFCC, other ministries, CPCB, and several SPCBs on the subject of co-processing.
- Capacity building of MoHUA, the state Urban Development Departments and Municipal Corporations on the subject of co-processing and utilization of RDF (GIZ, 2013).

- Organization of the stakeholder deliberations on the relevance of co-processing of wastes in the policy framework and sharing of best practices.
- Promoting the co-processing technology for the sustainable management of wastes (Dieter Mutz, 2014).
- Publication of results of the co-processing trial run for dry fractions of municipal solid waste with GIZ, Nashik Municipal Corporation, and geocycle in conferences and journals (Vaishali Nandan, 2014).
- Round table discussions were organized by MoHUA and GIZ for the preparation of the Municipal Solid Waste Management Manual 2016 with MoEFCC, CPCB, states, and cities as members of the expert committee deliberations with cement industries and other relevant stakeholders resulted in inclusion of co-processing of RDF in the Municipal Solid Waste Management Manual 2016 and the Solid Waste Management Rules 2016.

E. **SINTEF, Norway**

SINTEF is one of Europe's largest independent research organizations. Every year, several thousand projects are carried out by it for customers large and small spread out throughout the world.

Stakeholder awareness

Collaboration with CPCB and organizing workshops on AFR in different parts of the country for awareness generation.

- Publishing articles on the use of AFRs in scientific and other publications.
- Creating awareness among the authorities on the international best practices in co-processing.

Figure 15.8 depicts the photo of the workshops organized by CPCB and SINTEF for the promotion of co-processing in the country.

Advocacy

- Collaboration with CPCB in evaluating aspects related to co-processing for implementation in the policy framework.
- Capacity building of the authorities in MoEFCC and CPCB about co-processing.
- Collaborating with CPCB in drafting guidelines for co-processing and related matters.
- Advocating the benefits of managing the entire hazardous and non-hazardous wastes generated in Norway through co-processing in cement kilns.

F. **NCCBM**

National Council for Cement and Building Materials is an apex body dedicated to continuous research, technology development and transfer, and education and industrial services for the cement and building material industries in India.

Stakeholder Awareness

Fig. 15.8 Stakeholder workshops organized by CPCB, India, and SINTEF, Norway, in different cities in India. Dr. Kare Helge Karstensen, the author of the book, is seen along with CPCB and SPCB dignitaries

- Promoting use of AFRs in the cement industry by encouraging publications of scientific articles in the biennial NCB International Conference.
- Dissemination of knowledge in dedicated workshops/webinars (with participation of over 500) for executives and plant operators on aspects related to co-processing of AFR's.

Advocacy

- Promoting the advantages of co-processing among stakeholders and clarifying their concerns through research investigations, publications, demonstrations, etc.
- Formation of Expertise Group to address concerns of clinker manufacturing units related to process/operation, system design, cement chemistry, and environment
- For the first time in the AFR field, develop concepts for the co-processing of domestic and other wastes as alternative fuels, by checking the environmentally relevant and process-specific influences on the cement production. Accordingly, appropriate preparation and handling plants and dosing systems are planned for achieving the target %TSR.

- Supporting cement plants operating at low/marginal level of %TSR for enhancing the AFR consumption, without impacting the clinker quality, emission, and operation and achieving reduction in carbon footprint.
- National Council for Cement and Building Materials is an apex body dedicated to continuous research, technology development and transfer, education and industrial services for the cement and building material industries.

G. TERI

TERI is an independent, multi-dimensional organization, with capabilities in research, policy, consultancy, and implementation. It creates innovations and facilitates change management in the energy, environment, climate change, and sustainability space. It has pioneered conversations and action in these areas for over four decades and believes that resource efficiency and waste management are the keys to smart, sustainable, and inclusive development.

Stakeholder Awareness

- Organizing deliberations about co-processing of AFRs in the cement industry in its various stakeholder awareness initiatives.
- Promoting the concept of co-processing for sustainable management of solid wastes in its various projects.

Advocacy

- Promoting the advantages of co-processing among policy makers and other stakeholders.
- Preparation of Feasibility Study for RDF Availability for Cement Kilns in India to estimate the availability of RDF within 100, 200, and 300 km radius of cement kilns for promoting usage of municipal solid waste or MSW-based RDF for co-processing in cement plants in India and developing innovative business models to help facilitate this mechanism (TERI, 2017–18).
- Partnered with GIZ in assessment and mapping of cement plants and municipal solid waste processing facilities (TERI, 2017–18).

H. ISWMAW

International Society of Waste Management, Air and Water (ISWMAW) is a non-profit non-government organization, active in a variety of areas, including supporting collaborative research and publication, conferences, meetings, training programs, information development and dissemination, and technical assistance on a global scale having collaboration with more than 45 countries. ISWMAW promotes and develops global sustainable and professional waste management, Sustainable Development Goals (SDGs), and the transition to a Circular Economy. The international platform of ISWMAW for awareness generation and collaborative activities is

International Conference of Sustainable Waste Management and Circular Economy (IconSWM-CE). It has a significant impact on encouraging stakeholders to practice environmentally friendly activities and research for the last 12 years, focusing on policy instruments, resource efficiency, waste management, green manufacturing, inclusive sustainable development, recycling, co-processing, and other innovative technologies.

Stakeholder Awareness

- Instituted the "IconSWM-CE Excellence Award for waste co-processing in cement plants" for individuals and organizations having significant contributions to co-processing activities in cement plants including research since 2019. The processing of IconSWM-CE Excellence Award 2021 and 2022 for co-processing is under progress.
- Organizing deliberations about co-processing of AFRs in the cement industry in its various stakeholder awareness initiatives.
- Promoting the concept of co-processing for sustainable management of solid wastes in its various projects including research projects and publications.

To encourage coprocessing in the cement plants, ISWMAW has announced the participation in the competition for the "*IconSWM-CE Excellence Award for waste co-processing in cement plants, 2021*" in the 11th IconSWM-CE & IPLA Global Forum 2021 to be held during December 1–4, 2021.

Figure 15.9 provides photos of the the Award Ceremony in the inauguration of the 9th International Conference on Sustainable Waste Management and Circular Economy for presenting the "*IconSWM-CE Excellence Award for waste co-processing in cement plant, 2019*" by Prof. Sadhan Kumar Ghosh, President, ISWMAW.

Advocacy

- Promoting the advantages of co-processing among policy makers and other stakeholders.
- Research for the preparation of reports on corelation between co-processing and curbing marine littering in India and potential of MSW-based RDF for co-processing in cement plants in India during 2020–2022.

I.   Industrial Institute of Productivity India (IIPI)

Institute for Industrial Productivity India (IIPI) is an independent non-profit organization operating as a best practice network in partnership with government, industry, and community to develop and implement solutions that promote resource conservation, sustainable growth, and climate resilience. IIPI achieves this by driving demand for resource-efficient practices and clean technologies by undertaking action research, piloting new approaches, and developing and promoting innovative technologies. IIPI also develops tools and approaches that prepare stakeholders for a low-carbon future. To identify and resolve the regulatory and policy issues related to use of alternative fuels and raw materials (AFRs), a Forum of Regulators was created by IIPI with high-level representation from State Pollution Control Boards of major

Fig. 15.9 The ISWMAW instituted *"IconSWM-CE Excellence Award 2019"* for co-processing are being handed over to the organizations and individuals who contributed significantly to the promotion of co-processing in India. The awards were handed over by Prof. Sadhan K Ghosh, Chairman, IconSWM-CE, and President of ISWMAW, the author of the book, in the presence of the IconSWM-CE Co-chairs, Mr. Arne Ragossnig, President, ISWA, Austria, Dr. Abas Basir, DG, SACEP, Sri Lanka, Mr C. R. C. Mohanty, Env. Coordinator, UNCRD, Japan, Mr. K. Onogawa, Director, CCET-IGES, Japan, Dr. H. N. Chanakya, CST, IISc, the Vice chancellor, KIIT and the Jt. Secretary, UD, Government of Odisha

cement producing states in India. The forum met regularly to deliberate on the key policy and regulatory bottlenecks, and issued a series of five White Papers/Policy Briefs. These were developed in the following inputs from the members of state pollution control boards, technical experts, industry representatives and references to international best practices, and journals and research documents. Subsequently, the same has been published as a Compendium of five white papers on alternate fuel and raw (AFR) material use in cement manufacture. The compendium is the outcome of the forum's work to promote AFR use in the Indian cement industry.

J. Government Agencies

Co-processing being a sustainable initiative, this also was promoted by several government institutions. This included MoEFCC, CPCB, MoHUA, and several state pollution control boards that included Gujarat, Madhya Pradesh, Karnataka, Rajasthan, Chhattisgarh, Karnataka, Andhra Pradesh, Tamil Nadu, Telangana, etc. They actively granted the trials permission and also the regular permission for co-processing as per the defined protocols. They also encouraged the industry to accept co-processing as the preferred option for waste management. As co-processing option provides landfill-free status to the industries—which is an essential requirement for exporting products to many developing countries—DIPP also actively recommended the cause of co-processing to MoEFCC, CPCB, and MoHUA.

15.4.3   Notification of Rules, Emission Standards, Monitoring Protocol, and Guidelines

Based on the awareness initiatives and advocacy efforts implemented by different agencies, the Ministry of Environment, Forests and Climate Change (MoEFCC), Ministry of Housing and Urban Affairs, and Central Pollution Control Board released different notifications to facilitate co-processing in the country.

A.   Rules

The option of management of wastes using co-processing technology was included in the Hazardous and Other Waste Management Rules notified in 2016 (HOWM Rules, GOI, 2016). As per these rules, the mandatory requirement of the demonstration of the suitability of co-processing technology for the management of a given waste stream was removed. The new Rule 9 in the HOWM Rules 2016 notified that grant of permission for co-processing will be based on compliance with the notified Emission Standards for co-processing.

Co-processing in cement kilns was also incorporated in the Plastic Waste Management Rules 2016 (PWM Rules, GOI, 2016) for the management of non-recyclable plastic wastes. Subsequently, these rules have been amended as Plastic Waste Management (Amendment) Rules 2018.

Co-processing in cement kilns was further incorporated in the Solid Waste Management Rules 2016 (SWM Rules, GOI, 2016) for the sustainable management of segregated combustible fraction (SCF)—also termed as refuse derived fuel (RDF)—from municipal solid waste (MSW). The permission for co-processing of the plastic waste and SCF/RDF was also mandated based on compliance with the notified emission standards.

B.   Emission Standards

The Emission Standards for cement kilns undertaking co-processing of AFRs were notified in the Environment (Protection) Third Amendment Rules 2016 dated 10 May 2016 (MoEFCC—Emission Standards, 2016).

C.   Monitoring Protocol

The monitoring protocol for ascertaining compliance with the emission standards while undertaking co-processing was published by CPCB in 2018 (CPCB, 2018).

D.   Guidelines

The technical guidelines for pre-processing and co-processing of wastes were released by CPCB in July 2017 (CPCB, 2017). The guidelines for co-processing of plastic wastes were released by CPCB in May 2017 (CPCB, 2017). As per

Table 15.5 Growth of co-processing in India

Parameter	2008	2012	2016	2020
Number of plants having permission	< 10	~ 20	~ 45	> 60
Average TSR%	0.6%	~ 1.0	3.8%	> 4.0%

these guidelines, the permission for pre-processing and co-processing are being granted by SPCBs.

Considering the quality issue faced in co-processing of SCF/RDF from MSW in cement kilns, the Ministry of Housing and Urban Affairs (MoHUA) published guidelines on usage of refuse derived fuel in various industries in 2018 (MoHUA, 2018).

15.4.4 Status of Co-processing in India

The journey of the growth of co-processing in the Indian cement industry is represented in Table 15.5. This data is based on the assessment which was carried out by CII-Godrej GBC.

15.5 Case Studies of Cement Companies

The AFR journey is taken up very aggressively by several companies in India and the volumes in these companies are getting ramped up on a yearly basis. Case studies of a few of these companies are provided in the next pages for getting better clarity of this AFR journey of the Indian cement industry.

Case Study 1—ACC Limited and Ambuja Cement Ltd.

Case Study 2—JSW Cement Ltd.

Case Study 3—J K Lakshmi Cement Ltd.

Case Study 4—My Home Cement Ltd.

Case Study 5—Dalmia Cement Ltd.

Case Study 6—J. K. Cement Ltd.

Case Study 7—Vicat India.

It can be observed from these case studies that most of them are aligned to the Low-Carbon Technology Road map prepared for the Indian cement industry and are planning to achieve high level of TSR to reduce the carbon footprint drastically.

Case Study—1

Ambuja Cement Ltd & ACC Limited

In India, Ambuja Cements Ltd and ACC Ltd, which are a part of the global conglomerate LafargeHolcim, have been offering sustainable waste management solutions through co-processing, since 2005. The organizations are the pioneers of co-processing in the country and together have the capacity to manage more than 700,000 tons of waste through 13 co-processing and 6 pre-processing facilities. Plans are already in place to further enhance this capacity. Ambuja and ACC offer waste management solutions under the umbrella of Geocycle which is the global waste management brand of the LafargeHolcim Group. Globally, Geocycle treats around 10 million tons of waste annually, serving more than 10,000 customers in over 50 countries. It has access to more than 80 waste pre-treatment facilities and more than 180 cement plants, with dedicated co-processing installations. Over the years, the organizations put in concerted efforts in advocacy and capacity building to achieve greater technical and legal recognition for co-processing technology in line with accepted international standards. These efforts bore fruit when the changes in the Indian Waste Legislation in 2016, recognized co-processing as a preferred technology for waste management.

Leveraging the global experience and extensive know-how of their parent company, Ambuja and ACC have blazed a trail by supporting a widespread adoption of the co-processing technology in the country. This was done through extensive stakeholder engagement, collaboration with academia and scientific bodies and successful technical demonstration of the technology through more than 50 successful third-party emission monitoring trial burns. Under the umbrella Geocycle brand, the organizations have also been on the forefront of innovation, setting up several state-of-the-art facilities catering to varied waste streams, volumes and complexity. This has allowed them to venture into more complex waste streamsand facilitated many industries to achieve the landfill free status. This initiative also permitted sustainable solution for sorted municipal solid waste, thus supporting government initiatives like 'Swachh Bharat'.

Ambuja and ACC work with local communities to sustainably source agricultural waste (which can neither be fed to cattle or suitably used by agricultural producers) thus providing sustainable livelihood opportunities. In 2020, their activities under the aegis of Geocycle helped them source 100,000 tons of such agri-waste.

Their initiative has also been instrumental in remediation of 7 dumpsites and they are collaborating with more than 20 municipalities to support waste management in cities and towns across the country. The co-processing operations of Ambuja and ACC have consistently achieved a 15 to 20% growth in the last 3 years, treating more than 10,00,000 Tons of waste and alternative resources. Their endeavours have led to avoidance of more than 2,90,000 tons of waste going to landfills in 2020 alone.

www.acclimited.com

www.ambujacement.com

Case Study—2

Co-processing of alternative fuels by JSW Cement Limited

JSW Group, forayed into Cement market in the year 2009 with a vision to manufacture environment friendly construction materials utilizing industrial wastes and took initiatives with mission to curb climate change way ahead of everyone else. It positioned itself distinctively in a surplus supply industry by promoting Slag Cement & GGBS.

Today, JSW Cement Limited, has manufacturing plants at Vijayanagar in Karnataka, Nandyal in Andhra Pradesh, Dolvi in Maharashtra, Jajpur & Shiva Cement Limited (a subsidiary) in Odisha, Salboni in West Bengal and JSW Cement FZE, Fujairah (a subsidiary) in UAE with an annual installed capacity to produce 14 MTPA Cement products and GGBS, and is one of the largest companies in India to produce green construction products.

JSW Cement is operating an integrated cement plant at Nandyal which utilizes the latest energy efficient technology in clinker manufacturing process. As part of its sustainability initiative, it started co-processing of wastes (hazardous & non-hazardous) in kiln from FY-16 and has continuously increased the Thermal Substitution Rate (TSR) to 8.26% during FY-20. During this journey it has used multiple types of wastes such as pharmaceutical waste, liquid waste, plastic waste, carbon black, PPF oil, mixed industrial waste, dolo-char, biomass etc. and successfully co-processed these with minimal additional environment impacts. The company has co-processed ~88,500 MT of waste in cement kiln till FY-21, thus avoiding the usage of ~ 34436 MT of coal.

At its recently commissioned Fujairah clinker plant, it has co-processed Spent Pot liners (SPL), a waste from aluminum industry, which has given dual benefits i.e. improving TSR as well as reduction in specific thermal energy. A project is also underway to commence usage of RDF and is expected to be commissioned in FY22.

During our efforts in co-processing we have reduced ~ 75398 MT of CO_2 emissions till FY-21 by utilizing the wastes in place of coal. We have roadmap in place to take up our TSR rate to 30% by 2030 thereby making our manufacturing process more sustainable.

Benefits of co-processing include saving of fossil fuels; saving of land mass used for disposal of industrial waste and subsequent damage to environment & earth by industrial wastes. Apart from co-processing, we use industrial wastes or byproducts such as red mud, fly-ash, slag, chemical gypsum in the manu-facturing process, thus reducing the consumption of natural resources like limestone, bauxite, and mineral gypsum.

We have implemented various energy conservation/efficiency measures, and carried out third party product certifications like Environmental Product Decla-ration (EPD), Green-Pro for our PSC, CC & GGBS products. With the help of all these sustainable initiatives, our specific CO_2 emissions per MT of PSC (261 kg CO_2/MT of PSC) is one of least among the cement companies across the world.

www.jswcement.in or Email info@jswcement.in

Case Study—3

J K Lakshmi Cement Ltd.
(www.jklakshmicement.com)

JK Lakshmi Cement Limited (JKLC) is a part of the prestigious JK Organisation. This eminent industrial house is over a hundred and twenty-five years old and boasts operations in India and abroad with a leadership presence in the fields of tyre, cement, paper, power transmissions, sealing solutions, dairy products and textiles. Having started the company in 1982, we have modern and fully computerized, integrated cement plants at Jaykaypuram, in the Sirohi district of Rajasthan, at Dabok, in the Udaipur district of Rajasthan (a subsidiary of the company) and at Ahiwara, in the Durg district of Chhattisgarh. With the four split location grinding units, the combined capacity of our company is 13.3 Million MT per annum. Being a sustainability driven organisation, JKLC is at the forefront in utilising different kinds of wastes in the manufacture of cement. Apart from using substantial quantity of fly ash & Slag in the cement manufacture, we are also utilising large quantities of wastes generated from Agricultural, Municipal and Industrial sectors.

Our AFR journey started in year of 2005-06 while using of sludge from the textile industries, we are now associated with around more than 80 industries and providing them our co-processing solutions for sustainable management of the hazardous and non-hazardous wastes generated by them. Quite a few of them have attained the landfill free status. Our current TSR is 6 % which has helped us to reduce carbon foot print emission level by about 100,000 MT. The total quantum of AFR co-processed by us so far has been 5, 10,403 MT. This has helped us to reduce the 8, 00,000 Tons of CO_2 emission and Conserved about 3, 30,000 MT of fossil fuels. Through this initiative we have been able to provide direct employment to around 10 persons and indirect employment of 300 persons.

Today, our heads stand high in serving the nation. For the last 34 years we have been reinforcing the base with multiple superior quality products. The Company has maintained its status of being the least cost producer with increased customer satisfaction and loyalty. By implementing sustainable manufacturing practices across its different units, our company has very well carried out its duty towards promoting sustainable growth and development towards betterment of Environment, Community and Economy.

Shrimati Vinita Singhania—Vice Chairperson & Managing Director

Case Study—4

My Home Industries Private Limited, Mellacheruvu Cement Works, Telangana
(www.mahacement.com)

My Home Industries Private Limited, Mellacheruvu Cement Works (MCW), Telangana State was established during the year 1998 with an installed capacity of 0.2 MTPA and has now attained a capacity of 10 million tonnes per annum. MCW unit utilises liquid Pharma waste as alternative fuel in the kiln to reduce fossil fuel with an aim to avoid the environmental pollution, reduce the GHG emissions and conserve the fossil fuels and natural resources.

MCW has installed in 2012 a state-of-the-art closed circuit Alternative Fuel facility with unloading, storage and feeding system for co-processing liquid wastes from Pharmaceutical industries. This facility has been designed by giving due consideration to required safety standards for managing the hazardous wastes from pharmaceutical industry. The design of this facility takes care of the challenges associated with the pharmaceutical wastes such as low flash points, Chloride corrosion, and odours due to VOCs etc. This facility provides MCW an opportunity to achieve a TSR of 5% and the same is being planned for augmentation to achieve 10% TSR by 2030.

The total quantity of Liquid pharma waste co-processed in facility so far is more than 92,000 Tons, which has led to the conservation of more than 63,000 Tons of high-quality coal. This has also helped the company in reducing more than 1,01,000 T of GHG emissions . While implementing this co-processing initiative, MCW has helped more than 8 Pharmaceutical companies to manage their wastes in a sustainable manner. The direct employment generated due to this initiative is about 15 Nos. and has helped to provide indirect employment to > 60 persons. In the year 2019-20, more than 16,370 Tons of Pharmaceutical wastes were co-processed conserving about 11,000 Tons of coal and avoiding >18,000 Tons of GHG emissions.

The sustainability journey of MCW continues with higher strides adding value to environment and sustainability.

Case Study—5

Decoupling GHG emissions while following Clean and Green is Profitable & Sustainable

Dalmia Cement Bharat Limited (DCBL) is the fourth largest cement producer in India with a total installed capacity of 30.75 million tonnes. Dalmia Cement was the first in global heavy-industry sector to commit to 2040 carbon negative roadmap. The company was successful in visualising the future course of policy and actions in 2018 itself as more than 60% of the global GDP committed to Net Zero within two years. The carbon negative commitment of Dalmia Cement served as catalyst to change the image of heavy industry sector from Hard-to-Abate to Possible-to-Abate.

Mr. Mahendra Singhi, MD & CEO, Dalmia Cements Bharat Limited reiterated the commitment of at the United Nations Climate Ambition Summit 2020 which was participated by 70 national leaders. Highlighting time to act now due to unprecedented climate challenge being faced by the humanity. Mr. Singhi provided the details of DCBL climate actions towards more sustainable cement and construction sector. The company has a business philosophy of "Clean and Green is Profitable and Sustainable" which takes a co-benefit approach of profitability and sustainability. The company is globally ranked No 1 by CDP on business readiness for a low carbon economy transition. The group is nearly ten times water positive and is the first cement company in the world to join EP100, RE100 and EV 100 initiatives. Dalmia is also identified as Climate Defender by BBC World, Carbon Pricing Champion by the World Bank Group and COP-26 Business Leader by UN COP-26 Presidency.

Dalmia Group: Journey of Alternative Fuels:

Usage of Alternative fuel and material is one of the important lever to achieve Carbon negative commitment. Dalmia has advanced marching towards 2040 goal and made capital expenditure of more than 200 Crores towards required infrastructure for co processing. In FY 20-21 group kilns co processed more than 0.22 Million tonnes of hazardous and non-hazardous waste, major material being RDF, MLP and paper mill plastics. DCBL is a solid waste recycling positive company on account of its alternative fuels usage.

In FY20-21 average TSR for group cement kilns was more than 8%. DCBL is the first Indian Cement company to install Chlorine Bypass system to enhance usage of alternative fuels by end of year 2021 in two plants. Group has set an ambitious target of **achieving TSR of 35% in all group kilns by FY22-23 and achieving 100% fossil fuel replacement by 2035 through use of alternative fuels and sustainable biomass use**.

DCBL plants located in Meghalaya and Assam are consuming saw dust generated in saw mills and bamboo as a fuel which is grown locally. This initiative would gain further momentum considering Indian Government has declared Bamboo as a grass. Besides, nearly 26 million Ha. Wasteland identified by MoEFCC may be available for the sustainable biomass resource development and switching of fossil fuels.

Road Ahead:

Dalmia group is committed to its target of carbon negative roadmap & 100% switching of fossil fuels.
(www.dalmiacement.com)

Case Study—6

J K Cement Limited.
(www.jkcement.com)

JK Cement Ltd is one of India's leading manufacturers of Grey Cement and the third largest White Cement manufacturer in the World. Over four decades, the company has partnered India's multi-sectoral infrastructure needs on the strength of its product excellence, customer orientation and technology leadership. JK Cement's operations commenced with commercial production at its flagship grey cement unit at Nimbahera, Rajasthan in May 1975. The Company has an installed Grey Cement capacity of 14.7 MnTPA, making it one of the top cement manufacturers in the Country. JKCL is the No. 1 manufacturer of Wall Putty in the world with capacity of 1.2MnTPA and the third largest manufacturer of White Cement, globally, with total white cement capacity of 1.20 MnTPA.

JK Cement has always worked towards reducing carbon footprint and has made a shift to green-technology. In this drive JK Cement has been focussing on increasing the use of alternate fuels & raw materials and also setting up more solar, wind and waste heat recovery sources of energy generation. The company is utilising Solid and Liquid waste; combustible material from MSW, processed RDF, FMCG waste, AGRO waste & Industrial waste as an alternative fuel and GCP dust, Lime sludge, red mud and iron sludge as a Raw material in the kiln to reduce fossil fuel and natural raw material usage.

JK Cement has always worked towards reducing carbon footprint and has made a shift to green-technology. In this drive JK Cement has been focussing on increasing the use of alternate fuels & raw materials and also setting up more solar, wind and waste heat recovery sources of energy generation. The company is utilising Solid and Liquid waste; combustible material from MSW, processed RDF, FMCG waste, AGRO waste & Industrial waste as an alternative fuel and GCP dust, Lime sludge, red mud and iron sludge as a Raw material in the kiln to reduce fossil fuel and natural raw material usage.

The AFR Journey of JK Cement started in the year 2014 with 6000 MT consumption per annum and now the total quantity of waste co-processed in JK plants is more than 500,000 Tons which has led to the conservation of more than 125,000 Tons of high-quality coal. This has also helped the company in reducing >325,000 T of GHG emissions. JK Cement's Muddapur unit is running on +18% TSR and has facility of both Solid and Liquid feeding arrangement. In the year 2019-20, nearly 150,000 Tons of solid + liquid wastes were co-processed conserving about 80200 Tons of coal and almost 140,000 Tons of GHG emissions. In the year 2020-21, 205,000 Tons of solid + liquid wastes were co-processed conserving about 106,000 Tons of coal and reducing 180,000 Tons of GHG emissions.

JK Cement's endeavour is to create sustainable economic value for stakeholders while ensuring a positive impact around the communities they operate. JKCL has diligently crafted its sustainability strategy through an extensive process comprising internal and external consultations, peer benchmarking and its alignment with national and global goals. To further augment the same, company has built a solid governance ecosystem to ensure that efforts are carried out in a consistent, accountable and transparent manner.

Case Study—7

VICAT INDIA

Bharathi Cement Ltd. and Kalburgi Cement Ltd. are the flagship companies of Vicat, France in India. Vicat is the pioneer in cement manufacturing since its invention by Louis Vicat in 1817. Vicat, in India, operates with a name brand "Bharathi Cement" and "Vicat Cement" (only in Maharashtra). The company owns the state-of-the art manufacturing plants in Kadappa, Andhra Pradesh and Kalaburagi, Karnataka with a combined production capacity of 8.6 MTPA.

Since 2012, VICAT in India started replacing their traditional fuels with AFRs. For this, it has installed State-of-the-art co-processing facilities with stringent safety standards in both the plants to co-process many different kinds of hazardous and non-hazardous waste streams. These facilities provide Bharati Cement Ltd. and Kalburgi Cement Ltd. an opportunity to achieve a TSR of 19% and 23% respectively and is poised to achieve 35% TSR by 2030.

The total quantity of waste co-processed in these facilities annually is more than 2,25,000 Tons which has led to the conservation of around 1,40,000 Tons of high-quality coal. This has also helped the Vicat group to reduce >3,00,000 T of GHG emissions annually. The direct employment generated due to this initiative is 55 and has helped to provide indirect employment to >180 persons. While implementing this co-processing initiative, it has helped a large number of number of industries to ensure that their waste gets managed in a transparent, traceable and auditable manner.

To have sustainable utilization of waste, Vicat is extending its support and handholding with local waste processors by investing for pre-processing the waste in their premises. A recent tie up with Bhumi Greens in Pune is one of the benchmark example for the industry for such a collaboration. Along with M/s Bhumi Green Energy, Vicat India has invested for the 1000 TPD legacy municipal waste processing unit in Pune. The generated RDF is being transported and used at Vicat India Kalburgi Cement Plant in Karnataka. This MSW processing unit in running successfully and Vicat India is looking at several other cities to implement this kind of model for sustainable AFR production.

Vicat leads the country in co-processing drive with highest percentage of TSR.

(www.bharathicement.com)

15.6   Conclusion

After 2016, many cement plants have been observed to initiate co-processing and many who are already co-processing are able to ramp up their volumes due to facilitation that has happened in the policy framework and also the Swatch Bharat Abhiyan that has been initiated in the country from 2014. The major rise in volumes is from the Municipal sector on account of the reforms that have been happening towards the management of MSW and also due to remedying of the existing dump yards. Many cement plants have now > 10% TSR in their kilns and many of them are targeting bigger aspirations for 2030 and 2050 to move towards the net-zero ambition.

References

CII. (2012). *White Paper on Increasing Co-Processing in Cement Plants.* Shakti Sustainable Energy Foundation (shaktifoundation.in).

Circular from CPCB in Conjunction with Rule 9 of HWM Rules, 2016 with list of trialed waste streams, B-33014 /7/2016/PCI-II, Dated 30 Jun 2016

CPCB, Annual Report, 2003–2004. http://www.cpcbenvis.nic.in/annual_report/AnnualReport_8_AnnualReport_2003-04.pdf.

CPCB, Annual Report, 2005–2006. http://www.cpcbenvis.nic.in/annual_report/AnnualReport_21_AnnualReport_2005-2006.pdf.

CPCB, Annual Report 2004–2005. http://www.cpcbenvis.nic.in/annual_report/AnnualReport_21_AnnualReport_2004-2005.pdf.

CPCB. (2016). Guidelines for co-processing of plastic wastes in cement kilns as per Rule 5b of PWM Rules, May 2017.

CPCB. (2016). Guidelines for Pre-processing and Co-processing of Hazardous and Other Wastes in Cement Plant as per H & OW(M & TBM) Rules, July 2017.

CPCB. (2018). 1st Revised guidelines for Continuous Emission Monitoring System.

Eco-echoes. (2008). 9(4), 1–5, Eco-Oct-Dec-08.pmd (icpe.in)

Environment (Protection) Third Amendment Rules (2016). Notified by MoEFCC, Government of India (GSR497E) dated 10 May 2016, http://ismenvis.nic.in/Database/Notification_10th_May_2016-GSR497E_12779.aspx.

GIZ. (2018). Assessment and Mapping of Cement Plants, and Municipal Solid Waste (MSW) Processing, Facilities within 100, 200 and 300 km Radius, for Promoting RDF Co-processing in India.

GIZ. (2013). Status Report on Refuse Derived Fuel (RDF) utilisation in India.

GIZ/LafargeHolcim. (2018). Guidelines on pre-processing and co-processing of waste.

GTZ/Holcim. (2006). Guidelines on co-processing of wastes.

HOWM Rules. (2016). Notified by MoEFCC Government of India (GSR 395 E) dated 4 April 2016 available at http://moef.gov.in/wp-content/uploads/2017/08/GSR-395E.pdf

Karstensen, et al. (2014). Destruction of concentrated chlorofluorocarbons in India demonstrates an effective option to simultaneously curb climate change and ozone depletion. *Environmental Science & Policy, 38,* 237–244.

Mutz, D., & Nandan, V. (2014). Co-processing waste materials in cement production: Experience from the past and future perspectives. *International Journal of Environmental Technology and Management, 17*(2/3/4).

MoHUA, CPHEEO. (2018). Guidelines on usage of RDF in various industries, Oct available at 5bda791e5afb3SBMRDFBook.pdf (cpheeo.gov.in)

Nandan, V., Dube, R., Yadav, J., Parlikar, U., Nicole, G., Chakraborty, M., & Salunke, S. (2014). Co-processing trial of dry segregated rejects from Municipal Solid Waste in cement plants. In *Proceedings of the 4th International Conference on Solid Waste Management,* (pp. 376–388). Jan. 28–30.

Parlikar, U., Bundela, P. S., Baidya, R., Ghosh, S. K., & Ghosh, S. K. (2016). Effect of variation in the chemical constituents of wastes on the co-processing performance of the cement kilns. *Procedia Environmental Sciences, 35,* 506–512.

PWM (Amendment) Rules. (2018). notified by MoEFCC, Government of India (GSR 285 E) dated 27 March 2018 available at http://moef.gov.in/wp-content/uploads/2017/08/PWM-amendment-english-2018.pdf

PWM Rules. (2016). Notified by MoEFCC, Government of India (GSR 320 E) dated 18 March 2016 available at PWM-Rules-2016-English.pdf (moef.gov.in)

SWM Rules. (2016). Notified by MoEFCC, Government of India, (SO1357 E) dated 8 April 2016 available at http://moef.gov.in/wp-content/uploads/2017/08/SWM-2016-English.pdf

TERI, Annual report, 2017–2018, pp 33

The Hazardous Wastes (Management, Handling and Transboundary Movement) Rules. (2008). Government of India

WBCSD (2014). Cement sustainability initiative, guidelines for co-processing fuels and raw materials in cement manufacturing.

Appendix

Chapter 1

3R	Energy Intensive	POPs,
AFR	GHG emissions,	Polychlorinated dibenzofurans
Alternative Fuels,	Global CO_2,	(PCDFs),
Basel Convention,	Hazardous waste,	Primary fuel
Cement production	Incineration, The Combustion	Stockholm Convention,
Circular Economy,	efficiency,	Ton of oil equivalent (TOE),
Co-processing,	Indian Cement Industry	Transforming Our World,
Earth Summit,	Landfill,	Trial burn,
Electrical energy consumption	Our Common Future,	WCED,
	Polychlorinated	World Conservation Strategy,
	dibenzo-p-dioxins (PCDDs),	

S. K. Ghosh et al., *Sustainable Management of Wastes Through Co-processing*,
https://doi.org/10.1007/978-981-16-6073-3

Chapter 2

Alternate use	Dust	Plastic waste
Alternative fuels and raw materials (AFR)	Environmentally sound management	Pre-calciner
Alternative raw materials	Environmentally sound technologies	Preheater
Authorisation		Pre-processing
Auto-Ignition Temperature	Extended Producer Responsibility (EPR)	Producer
Basel Convention		Recovery
Biodegradable waste	Facility	Recycling
Bio-methanation	Flash Point	Refuse Derived Fuel (RDF)
Brand owner	Food-stuffs	Registration
Calcination	Fuels	Prescribed Authority
Captive treatment, storage and disposal facility	Handling	Reuse
Carry bags	Heating (calorific) value	Rotary kiln
Cement	Higher heating (calorific) value (HHV)	Sanitary land filling
Central Pollution Control Board (CPCB)	Importer	Segregation
Clinker	Incineration	Solid waste
Clinkering	Inert	Solid Waste Management (SWM)
Co-incineration plant	Institutional waste generator	Sorting
Combustible waste	Kiln	Storage
Commodity	Leachate	State Pollution Control Board
Common treatment, storage and disposal facility	Local Body	Street vendor
Composting	Lower heating (calorific) value (LHV)	Sustainable development
Concrete	Manufacturer	Tipping (or Gate) Fee
Consent	Multi-layered packaging	Transboundary movement
Contractor	Non-biodegradable waste	Transportation
Co-processing	Non-Recyclable Plastics Waste (NRPW)	Transporter
De-centralised processing	Operator of a facility	Treatment
Disintegration	Operator of a facility	User fee
Disposal	Other wastes	Utilisation
Dry waste	Plastic	Virgin plastic
Dump sites	Plastic sheet	Waste
		Waste generator
		Waste pickers

Chapter 3

Chapter 4

Alternative Fuels	Clinkerization	Aluminium
Cement	Fuel preparation	Bag filters
Clay	Kiln feed	By-pass
Coal	Petcoke	Calcium
Dispatch	Pozzolanic materials	Dry process
Fuel preparation	Raw material preparation	ESP
Grinding	RDF	Exhaust gas
Limestone	Rotary kiln	Flame temperature
Manufacturing process	Semi dry process	Iron
Mines	Semi wet process	Raw mix
Portland cement	VSK	Retention time
Quarrying	Wet process	Silica
Raw materials preparation		

Chapter 5

Raw mix	Silica	PSC
Coprocessing	Alumina	Fly ash
AFRs,	Iron	GBFS
Cement manufacture	OPC	Limestone
Cement chemistry	PPC	Standards
Operations	Rotary kiln	Clinker phases
Quality control	Compressive strength	Degre of calcination
Calcium	Flame photometer	Loss on Ignition
C3S	Gravimetric analysis	% Liquid
C2S	Volumetric	Sulfur to Alkali Ratio
C3A	Complexometric	Free Lime
C4AF	Raw meal	Excess sulphur
Ash	Alumina Modulus	Blending Ratio
XRF	Silica Modulus	Kiln feed to clinker factor
XRD	Raw meal to clinker ratio	Clinker to cement factor
Optical Microscopy	LSF	Insoluble residue
Coal ash	Hydraulic Modulus	Volumetric loading
Burner	Bogue's Formulae	Thermal loading
Theoretical air	% Excess Air	Feed moisture evaporation rate
Primary air momentum	GCV	False air
Burner nozzle velocity	NCV	
Position of burner	Ultimate Analysis	
	Proximate Analysis	

Chapter 6

Waste Acceptance	Bomb Calorimeter	MSDS
Waste evaluations	Flash Point analyser	Inspection
Waste Handling	Chloride titrator	Radioactive wastes
Transport	Sulphur analyser	Segregation
Non-compliant deliveries	GC MS	Design considerations
Labelling	ICP AES	VOC
Types of AFRs	Stack	House keeping
Pre-processing	Emission monitoring	Storage time
Storage	Guidelines	Fire detection and control
BAT	Spills	PCDD
BEP	Safety	PCDF
Quality	EIA	AFRs
Emission monitoring	Calibration	Input control
UN Dangerous Goods labelling	Trial	Feed point
Compatibility of chemicals	Training	Main burner
Hazardous wastes	EMS	Kiln inlet
Pre-processing	Banned wastes	Raw mill
Co-processing	Risky wastes	Process control
CEMS	Acceptance control	Laboratory
Stack monitoring	Negative list	Limit values
Test Burn	PPE	POHC
	DE	DRE
	POPs	

Chapter 7

Sustainability	Water	International Energy Association
Cement manufacture	Plastics	(IEA)
Co-processing	Packaging	Technology Road-map
Natural Capital	Acidic emissions	Primary emissions
Carbon foot-print	Profits	Secondary Emissions
Emissions	Cement Sustainability Initiative	Tertiary emissions
GHGs	(CSI)	CO2
Particulate matter	WBCSD	Climate Change
SOx	Battelle Memorial Institute	Energy Efficiency Improvement
NOx	Agenda for Action	Clinker substitution
VOC,	Resource efficiency	Carbon capture and utilization
Mining	Conservation	(CCU)
Bio-diversity	Low carbon road map	Carbon capture and Storage (CCS)
Economics	Acid due point	Renewable energy
Circular economy	Rankine cycle	LC3 Cement
Lever	Organic Rankine cycle	FGD Technology
Fly ash	Kalina cycle	EBA Process
GBFS	Geopolymer	Water conservation
Volcanic ash	Algal growth	Harvesting
Waste heat recovery	Sustainable mining	Plastic
Water table	Challenges	
Plantation	Packaging	
TSR		

Chapter 8

Environmental regulation	Japan	Solid Waste Management
Pollution	South Africa	Rules
Policy	River	NIMBY
Act	Valley	Permit
Law	Framework	SOP
Rule	Factories Act	Single Window
Guidelines	Water Act	EPR
Technology	Air Act	GTZ
Co-processing	Environment Protection Act	Holcim
Pre-processing	Motor Vehicles Act	Technical Guidelines
Constitution	Public Liability Insurance Act	Basal
State Government	National Environmental	LafargeHolcim
Central Government	Tribunals Act	WBCSD
India	National Environmental	GIZ
Brazil	Appellate Authority Act	UNEP
USA	Labour	SINTEF
European Union	DGFASLI	ADEME
Germany	CPCB	Ecology
MoEFCC	SPCB	Emission Limits
Hazardous & Other Waste	Forest Conservation Act	WID
Management Ruels	EPA	BAT
Plastic Waste MAnageemnt	Clean Air Act	WFD
Rules	Eco Towns	EC
Industry Green Development	MACT	IPPC
Plan	CISWI	Denmark
NSR / PSD	Ireland	Hungary
NDRC		Netherlands
UK		Czech Republic

Chapter 9

Emission	Raw materials	CEMS
Considerations	Kiln dust	CPCB
Thermal treatment	Kiln system	SPCB
Emission Impact	Flora	NO
Co-processing	Fauna	NO_2
Standards	Diseases	P_2O_5
Particulate emissions	Gravity settling	CO_2
Acidic emissions	Mechanical collectors	Thermal NOx
Dust emissions	Wet scrubbers	Pyritic sulphur
Stack emissions	ESP	DeNOx
Fugitive emissions	Fabric filters	DRE
PM	NOx,	Opacity
FGD	SOx,	FTIR
SCR	H_2S	VOC
SNCR	SO3	Hg
IPCC	SO_2	Dioxin & Furan
WBCSD	HCl	Toxic
CCS	HF	AFR
Clinker	Kiln end	Kiln Burner

Chapter 10

Co-processing	SWM Rules	Dioxin
Thermal Treatment	HOWM Rules	Furan
RII	PWM Rules	Heavy Metals
AFRs	SPCB	Waste to Energy
Clinker	ODS	Municipal
Fossil fuels	Principles of co-processing	RDF
Wet Process	Quality	SCF
ILC	Laboratory	Dried Sewage Sludge
SLC	Ultimate Analysis	Industrial
Hazardous Waste	Proximate Analysis	Non-hazardous
Pre-processing	Gas Chromatograph	Agricultural
TSR	Finger Print	Agro-waste
Flash Point	Gross Calorific Value	Banned
Auto Ignition Temperature	Pollution	Radio active
MSDS	Spill	Bio-medical
Sustainable Waste Management	Concrete floor	Asbestos
Lumpy Waste	Odour	Electronic
Coarse Waste	Fragrance	POPs
Fine solid waste	Activated Carbon	Batteries
Liquid Waste	Polymer sheets	Basel
Zero Waste	Zeolite	CPCB
Emissions	Incompatible materials	Plastic
Chlorine	Electrical installations	Rubber
Fluorine	Chemicals	Net Calorific Value
Sulphur	Fire Fighting	Ash
Pre-calciner	AFR Feeding	Moisture
Kiln inlet	Main burner	Viscosity
Mid kiln	Shut Off Gate	CEMS
Transportation	Tyres	SOx
Lift	Tyre chips	NOx
Winch	Statutory	HCL
Conveyor	Liability	HF
Calciner Floor	Sustainability	VOC
Sludge	Waste Management Hierarchy	NH3
Feeding systems	Specific Thermal Energy	Dust
Double Flap Valve	Consumption	Screw Conveyor
		Rotary Air Lock
		Coating

Chapter 11

Pre-processing	Entry gate	Environmental provisions
Co-processing	Weigh Bridge	Safety provisions
Wastes	Laboratory	VOC emissions
AFRs	Solid wastes	Floor spillage
MSW	Liquid Waste	Odor control
Recyclable waste	Sludge waste	Safety provisions
Calorific Value	Shredder	Toxic materials
Cement kiln	Screens	Runaway reaction
Alternative Fuel	Shaft design	Fire prevention
Quality of clinker	Hammer mill	Rolling Bed Dryer
Unit operations	Grinder	Conduction Dryer
Blending	Granulator	Convection Dryer
Shredding	Moisture	Sludge
Drying	Thermal Energy	RDF
Impregnation	Fluidised bed dryer	Segregation
Size separation	Spray Drier	Bailing
	Rotary Dryer	

Chapter 12

Co-processing	Manpower	Fuel mix
Operation	Manual	AFR mix
Sustainability	SOP	MSDS
Legal	Emergency management	HAZOP
Environmental	External communication	PPE
Technical	Design considerations	Emergency Response Plan
Quality	Spillage	Geopolymer
Health	Fire	Burner Momentum
Safety	Odor	Odor control
Acceptance	Dust	Misting Systems
Sampling	Laboratory	Catalytic oxidation
Analysis	CEMS	Thermal Oxidation
Chlorine	TSR%	Zeolite based curtain
Sulphur	A/S ratio	Hammers
Alkali	Chlorine bypass	Rings
Raw mix design	Build-up	
Coating	Chlorin limits	

Chapter 13

AFR	Minimum Acceptable Price	Gate Fee
Alternative Fuels and Raw Materials	Mechanized	Operation and Management
Economic Parameters	Manufacture	Saving
Electricity Cost	Mill scale	SCF
Elementary Incineration	Raw material	Solid AFR
Investment	Resource	Salient features
Business Principle	RDF	Solid
Business Model	Raw meal	Hazardous
Biomass	Negative Variable	Landfill
Fixed Cost	Natural	Liquid AFR
Fuel cost	Non-Hazardous	Waste
Cost	Variable Cost	Pre-processing
Clinker	Viability	Transport Cost
Co-processing	Zero Landfill	Tipping Fee
	Challenges	TSR%

Chapter 14

Resource Intensive	Emissions	Alternative Fuels
Fossil fuels	Power	AF
CO2	Heat	TSR%
Sustainable	Minerals	Sustainability
Co-processing	AFR	GNR
Carbon foot-print	Biomass	Cement production
GCCA		

Chapter 15

Stakeholder	External Affairs	Brochures
Engagement	Impact	Flyers
Co-processing	Messages	Leaflets
Authorities	Communication plan	Cooperative Federation
Communities	Assessment	Community Advisory Panels
Activists	CSR	Mission
NGOs	Philanthropic	Mapping
Associations	Human right	Trust
Permits	Advocacy	LafargeHolcim
WBCSD	Ambuja	Guidelines
Co-processing	Ultratech	Mapping
GCCA	JK Lakshmi	SINTEF
CPCB	Geocycle	NCCBM
Pre Trial	Stakeholder meets	TERI
Post Trial	Plastic waste	Monitoring Protocoal
Trial	Variation in constituents	Status
Emission monitoring schedule	POPs	SSEF
Co-incineration	ODS	Geocycle India
ETP sludge	DRE	GTZ,
Chemical Industry	Awareness	GIZ
Paint Sludge	Advocacy	CETP sludge
Tyre chips	CII	Holcim
TDI Tar	Godrej-GBC	ACC
	CMA	

Bibliography

Agarwal, S. K. et al., (2001). Use of de-carbonated material 'LD slag' in the manufacture of portland clinker. In *Proceedings of 16th NCB International Seminar.*

Baidya, R., Ghosh, S. K., & Parlikar, U. V. (2016). Co-processing of industrial waste in cement kiln–a robust system for material and energy recovery. *Procedia Environmental Sciences, 31,* 309–317.

Baidya, R., Ghosh, S. K., Ghosh, S. K., & Parlikar, U. (2015). Co-processing of waste in cement plants. In *Proceedings of the International Conference on Solid Waste Technology and Management* (pp. 513–525).

Baidya, R., Ghosh, S. K., & Parlikar, U. V. (2015). Co-processing of waste—a robust process of energy utilisation from industrial wastes. In *Proceedings of the International Conference on Waste Management and Technology* (pp. 317–325) (Excellent Paper Award).

Baidya, R., Ghosh, S. K., & Parlikar, U. V. (2015). Sustainability of cement kiln co-processing of wastes in India: A pilot study. In *Proceedings of the International Conference on Solid Wastes 2015* (pp. 843–846). Knowledge Transfer for Sustainable Resource Management, ISBN 978-988-19988-9-7.

Baidya, R., Ghosh, S. K., & Parlikar, U. (2016). Influence of varying waste characteristics on kiln emission. In *Proceedings of the Asia-Pacific Conference on Biotechnology for Waste Conversion.*

Baidya, R., Ghosh, S. K., & Parlikar, U. V. (2017). Sustainability of cement kiln co-processing of wastes in India: A pilot study. *Environmental Technology, 28,* 1–10. https://doi.org/10.1080/095 93330.2017.1293738.

Baidya, R., Ghosh, S. K., & Parlikar, U. (2017). Co-processing of marble slurry in Indian cement plant. In *Proceedings of the 4th 3R International Scientific Conference on Material Cycles and Waste Management.*

Bundela, P. S., Parlikar, U., & Gautam, S. P. (2011). Co-processing of oils rags waste from Ford India at ACC Madukkarai. *Journal of Applied Sciences in Environmental Sanitation, 6*(3), 343.

Bundela, P. S., Parlikar, U., & Gautam S. P. (2011). Co-processing of ETP Sludge at ACC Madukkarai. *Recent Research in Science and Technology, 3*(9).

Bundela, S. P., Ulhas, P., & Milind, M. (2014). From Grey to Green : Waste Co-processing in Cement kilns. In *Proceedings of the 4th International Conference on Solid Waste Management* (pp. 347–352), 28–30 Jan 2014.

Bundela, P. S., Chakrawarty, M., & Gautam, S. P. (2010). Co-processing trial of spent carbon at Wadi Cement Works Karnataka. *American Journal of Environmental Sciences, 6*(4), 371–378.

Gautam, S. P., Bundela, P. S., & Chawla, V. (2009). Co-processing of plastic waste with coal in the cement Kiln. In *Proceedings of the Journal of Solid waste Technology Management 24th Conference.*

© The Editor(s) (if applicable) and The Author(s), under exclusive license to Springer Nature Singapore Pte Ltd. 2022
S. K. Ghosh et al., *Sustainable Management of Wastes Through Co-processing,*
https://doi.org/10.1007/978-981-16-6073-3

Gautam, S. P., Mohapatra, B. N. et al., (2009). Energy recovery from solid waste in cement rotary kiln and its environmental impact, Published and Presented In *Proceedings of 24th International Conference on Solid Waste Technology & Management* (pp. 15–18).

IIP / CMA (2015). Financial viability and barriers for implementing AFR project—an industry perspective. In *Proceedings of the 2nd International Conference on Enhancing use of AFR in the Indian Cement Industry*, Feb 19–20.

Kamyotra, J. S., Bala S. S., Gupta, P. K., Helge, K. K., & Parlikar, U. V. (2013). Cement kiln co-processing technology for management of waste from automobile industry in an ecologically sustaining Manner. In *Proceedings of the 13th NCB International Seminar on Cement and Building Materials*.

Liju, V. et al. (2019). Investigation on utilization of wollastonite in manufacture of OPC clinker. In *Proceedings of 16th NCB International Seminar*.

Mohapatra, B. N. et al. (2012). Co-processing plastic waste. *Indian Cement Review*, 42–45.

Mohapatra, B. N. et al. (2012). Energy recovery & eco friendly disposal of plastic waste by co-processing in cement kiln. *CMA Quarterly Journal Cement, Energy & Environment*, *11*(3).

Mohapatra, B. N. et al. (2014). Indian experience in using AFR in cement kiln industrial angles, *3*, 7–13.

Mohapatra, B. N. et al. (2019). Use of alternative fuels and raw materials in cement industry in India—prospects and challenges. In *Proceedings of CMA Conserve Conference on 30th Sep*.

Mohapatra, B. N., James, S., & Gupta, R. M. (2009). Successful utilization of processed municipal solid waste (RDF) as an alternate fuel in cement kiln. In *Proceedings of the Presentation & publication in 11th NCB International Seminar 2009 from 17th–20th Nov*.

Mohapatra, B. N. et al. (2014). Successful co-processing of hazardous waste, opium marc in cement kiln. In *Proceedings of the 29th International Conference on Solid Waste Technology & Management in March 30-April 2*.

Nandan, V., Dube, R., Yadav, J., Pawar, R., & Parlikar, U. (2014) Co-processing trial run with pre-processed Municipal Solid Waste Fractions. *ZKG International*, (7–8) AFR Special.

Parlikar, U., Bundela, P. S., Baidya, R., Ghosh, S. K., & Ghosh, S. K. (2016). Effect of variation in the chemical constituents of wastes on the co-processing performance of the cement kilns. *Procedia Environmental Sciences, 35*, 506–512.

Parlikar, U. (2015). Managing hazardous and other wastes through cement kiln co-processing in a sustainable manner, IGCW 2015, Dec 4–5

Parlikar, U. (2014). Technological advancements for manufacturing cement in an ecologically sustaining manner in the future. *Indian Concrete Journal*.

Parlikar, U., Ahuja, D., Sengupta, B. S., Baidya, R., & Ghosh, S. K., (2015). Disposal of wastes by co-processing in cement kilns—influence of their varying characteristics on critical kiln emission parameters. In *Proceedings of the 14th NCB International Seminar on Cement and Building Materials. (Excellent Paper Award)*.

Parlikar, U., Goyal, A., Mishra, A., Shah, M., Hill, A., & Ghosh, S. K. (2015). Evaluation of Use of SRF as AFR in Cement kiln, IconSWM.

Parlikar, U. (2015). Managing hazardous and other wastes through cement kiln co-processing in a sustainable manner. In *IGCW 2015*, 4–5 Dec 2015. Mumbai, India

Parlikar, U. (2017). Recommended quality standards for SRF/RDF derived from MSW to improve their resource recovery in cement kilns, Special Publication of Cement Manufacturers' Association on Alternative Fuels and Raw Materials. In *Proceedings of 3rd International Conference on AFR* (pp. 23–24).

Parlikar, U., Kiran Ananth, P. V., Muralikrishnan, K., & Kannan, V. (2017). Waste mapping and forecasting for alternative fuel usage in cement plants, Special Publication of Cement Manufacturers' Association on Alternative Fuels and Raw Materials. In *Proceedings of 3rd International Conference on AFR* (pp. 23–24).

Yadav, D. et al. (2019). Utilization of leather sludge in cement manufacture. In *Proceedings of the 16th NCB International Seminar*.

Printed in Great Britain
by Amazon

44192343R00231